FINITE ELEMENT METHODS FOR FLUIDS

FINITE ELEMENT METHODS FOR FLUIDS

Olivier Pironneau

Université Pierre-et-Marie-Curie & INRIA, France

JOHN WILEY & SONS
Chichester   New York
Brisbane   Toronto   Singapore

MASSON
Paris   Milan
Barcelone   Mexico

1989

Published with the support of
the Ministère de la Recherche et de la Technologie - DIST (France)

This book appears in French in the series
"Recherches en Mathématiques Appliquées"
edited by P.G. Ciarlet and J.-L. Lions (1988).

Copyright © Masson, Paris, 1989

ISBN 0 471 92255 2 (Wiley)
ISBN 2 225 81863 0 (Masson)

Library of Congress Cataloguing in Publication Data
Pironneau, Olivier
Finite element methods for fluids.
  Bibliography: p. 184
  Includes index.
  1. Fluid mechanics.   2. Finite element method.
I. Title.
TA357.P625 1989   620.1'06   89-9157
ISBN 0 471 92255 2

British Library Cataloguing in Publication Data
Pironneau, Olivier
Finite element methods for fluids.
  1. Fluids. Dynamics, Mathematics. Finite element methods.
I. Title.
532'.05'01515353

Printed and bound in France

Table of contents

Foreword

This book is written from the notes of a course given by the author at the Université Pierre et Marie Curie (Paris 6) in 1985, 86 and 87 at the Master level. This course addresses students having a good knowledge of basic numerical analysis, a general idea about variational techniques and finite element methods for partial differential equations and if possible a little knowledge of fluid mechanics; its purpose is to prepare them to do research in numerical analysis applied to problems in fluid mechanics. Such research starts, very often with practical training in a laboratory; this book is therefore chiefly oriented towards the production of programs; in other words its aim is to give the reader the basic knowledge about the formulation, the methods to analyze and to resolve the problems of fluid mechanics with a view to simulating them numerically on computer; at the same time the most important error estimates available are given.

Unfortunately, the field of Computational Fluid Dynamics has become a vast area and each chapter of this book alone could be made a separate Masters course: so it became necessary to restrict the discussion to the techniques which are used in the laboratories known to the author, i.e. INRIA, Dassault Industries, LNH/EDF-Chatou, ONERA...Moreover, the selected methods are a reflection of his career (and his antiquated notions ?). Briefly, in a word, this book is not a review of all existing methods.

In spite of this incompleteness the author wishes to thank his close collaborators whose work has been presented in this book and hopes that it would not hurt them to have their results appear in so partial a work: MM C. Bernardi, J.A. Désidéri, F. Eldabaghi, S. Gallic, V. Girault, R. Glowinski, F. Hecht, C. Parès, J. Périaux, P.A. Raviart, Ph. Rostand, and J.H. Saiac. The author wishes to thanks his colleagues too, whose figures are reproduced here: Mrs C. Begue and M.O. Bristeau, MM B. Cardot, J.M. Dupuy, F. Eldabaghi, J. Hasbani, F. Hecht, B. Mantel, C. Parès, Ph. Rostand, J.H. Saiac,V. Schoen and B. Stoufflet as well as the following institutions AMD-BA, PSA, SNCF and STCAN with special thanks to AMD-BA for some of the color pictures. The author also wishes to thank warmly MM P.G. Ciarlet and J.L. Lions for their guidance for the publication of this work, G. Arumugam and R. Knowles for their assistance in the translation into English and finally Mrs C. Demars for typing the script on MacWrite^{TM}. This book has been typeset with TEX; the translation from MacWrite to TeXture^{TM} has been made with EasyTeX^{TM}.

Notation

References:
A number in brackets, ex (1), refers to the equation (1) of the current chapter . The notation (1.1) refers to the equation (1) of the chapter 1.

The numbers in square brackets like [2] indicate a reference in the bibliography.

Vectors, matrices, scalar products ...
Unless stated otherwise the repeated indices are summed. Thus,

$$u.v = u_i v_i$$

denotes a scalar product of u with v.

For all vectors u, v and for all tensors of order 2 A, B we denote :

$$(u \otimes v)_{ij} = u_i v_j; \quad trA = A_{ii}; \quad A : B = tr(AB^T), \quad |A| = (trAA^T)^{\frac{1}{2}}$$

Differential operators :
The partial derivatives with respect to t or x_j are denoted respectively

$$v_{,t} = \frac{\partial v}{\partial t} \quad v_{i,j} = \frac{\partial v_i}{\partial x_j}; \quad v_{,j} = \frac{\partial v}{\partial x_j}$$

The classical grad, div, curl and laplacian operators are denoted respectively

$$\nabla p, \quad \nabla.v, \quad \nabla \times v, \quad \Delta v$$

The operator ∇ (nabla) applies also to vectors and tensors ... for example

$$\nabla . A = A_{ij,i} \quad (\nabla v)_{ij} = v_{j,i} \ (\text{ and thus } u.\nabla v = u_i v_{,i})$$

Little functions:
$O(x)$: Any function obeying $||O(x)|| \leq c$;
$o(x)$: Any function obeying $\lim_{x \to 0} ||o(x)||/||x|| = 0$.

Geometry :
The domain of a PDE is in general denoted by Ω, which is a **bounded open set** in R^n (n=2 or 3); its volume element by dx and the boundary by $\partial\Omega$ or often Γ . The domain is, in general sufficiently regular (locally on the one side of the boundary which is Lipschitz continuous) for us to define a normal n(x) for almost all x of the boundary Γ, but we admit domains with corners. The tangents are denoted by τ in 2D and τ_1, τ_2 in 3D; the curvilinear abscissa is noted by s and the element length (or area) is denoted by $d\gamma$.
$|\Omega|$ stands for the area or volume of Ω.
$diam(\Omega)$ is its diameter .

Function Spaces
P^k = space of polynomials of degree less than or equal to k.
$C^0(\Omega)$ = space of continuous functions on Ω.
$L^2(\Omega)$ = space of square integrable functions. We note the scalar product of L^2 by $(,)$ and the norm by $| \ |_0$ or $| \ |_{0,\Omega}$:

$$(a, b) = \int_\Omega a(x).b(x); \quad | \ |_0 = (a, a)^{\frac{1}{2}}$$

$H^1(\Omega)$ = Sobolev space of order 1.We note its norm by $|| \ ||_1$ or $|| \ ||_{1,\Omega}$.
$H_0^1(\Omega)$ = spaces of $H^1(\Omega)$ functions with zero trace on Γ.
$H^k(\Omega)^n$ = Sobolev space of order k with functions having range in R^n.

Finite Elements:
A *triangulation* of Ω is a covering by disjoint triangles (tetrahedra in 3D) such that the vertices of $\partial\Omega_h$, the boundary of the union of elements, are on the boundary of Ω. The singular points of $\partial\Omega$ must be vertices of $\partial\Omega_h$. A triangulation is *regular* if no angle tends to 0 or π when the element size h tends to 0. A triangulation is *uniform* if all triangles are equal.

An *interpolate* of a function φ in a polynomial space is the polynomial function φ_h which is equal to φ on some points of the domain (called *nodes*).

A finite element is *conforming* if the space of approximation is included in H^1 (in practice continuous functions). The most frequently used finite element in this book is the P^1 conforming element on triangles (tetrahedra).   An element is said to be *lagrangian* (others may be Hermite..) if it uses only values of functions at nodes and no derivatives.

Introduction

Computational fluid dynamics (CFD) is in a fair way to becoming an important engineering tool like wind tunnels. For Dassault industries, 1986 was the year when the numerical budget overtook the budget for experimentation in wind tunnels. In other domains, like nuclear security and aerospace, experiments are difficult if not impossible to make. Besides, the liquid state being one of the four fundamental states of matter, it is evident that the practical applications are uncountable and range from micro-biology to the formation of stars. At the time of writing, CFD is the privilege of a few but it is not hard to foresee the days when multipurpose fluid mechanics software will be available to non-numerical engineers on workstations.

Simulations of fluid flows began in the early 60's with potential flows, first incompressible or compressible hypersonic, then compressible transonic (cf. Ritchmeyer - Morton [1]). The calculations were done using finite difference or panel methods (cf Brebbia[2]). Aeronautic and nuclear industries have been the principal users of numerical calculations. The 70's saw the first implementation of the finite element method for the potential equation and the Navier-Stokes equations (Chung[],Temam[3], Thomasset[4]); also during that time the development of finite difference methods for complex problems like compressible Navier-Stokes equations continued (see Peyret-Taylor [5] for example). Recent years have seen the development of faster algorithms for the treatment of 3D flows (multigrid, (Brandt[6], Hackbush[7]), domain decomposition (Glowinski [7]), vectorization (Woodward et al[8]),the development of specialized methods to reach certain objectives (spectral methods, (Orszag [9]), particle methods (Chorin [10]) methods of lattice gas (Frisch et al [11]) for example) and the treatment of problems which are more and more complex like the compressible Navier-Stokes equations with interaction of shock-boundary layers, Knutsen boundary layers (rarefied gases, see Brun[ ]), free surfaces ...And yet in spite of the apparent success the day to day problems like flow in a pipe, in a glass of water with ice, in a river, around a car still remain unsolvable with today's computers. Simulation of turbulence, Rayley-Benard instabilities ... lie still further in the future. Suffice it to say that the subject will have to be studied for a good number of years yet to come!

Why restrict ourselves only to finite element methods ? many reasons; the first because one can not explain everything in 200 pages, the next because the author has more experience with this method than with other methods but most importantly because if one needs to know only one method then this is the best one to know ; in fact, this is the only technique (in 1987) which can be applied to all equations without restriction on the domain occupied by the fluid.

We have also chosen to study low degree finite elements because practical flows are often singular (shock, turbulence, boundary layer) and to do better with higher order elements one has to do a lot more than just doubling the degree of approximation (Babuska[], Patera[])

The chapters of this book more or less reflect the historical discovery of the methods. But since this book addresses itself to specialists in numerical analysis, *chapter 1* recalls the main equations of fluid mechanics and their derivations.

Chapter 2 treats of irrotational flow, i.e. the velocity of the flow is given by the relation

$$\mathbf{u} = \nabla\varphi \qquad (1)$$

where φ , the flow potential, is a real valued function .

Incompressible potential flows provide an occasion for us to recall the iterative methods used to solve Neumann problems by the finite element method:

$$-\Delta\varphi = 0 \text{ in } \Omega; \qquad \frac{\partial\phi}{\partial n} = 0 \text{ on } \Gamma \qquad (2)$$

The *transonic potential* equation is also treated in Chapter 1 :

$$\nabla.[(1 - |\nabla\varphi|^2)^{\frac{1}{\gamma-1}}\nabla\varphi] = 0 \text{ in } \Omega; \qquad \frac{\partial\phi}{\partial n} = 0 \text{ on } \Gamma \qquad (3)$$

and solved by an optimization method.

This chapter ends with the description of a potential correction method for weakly irrotational flows. So it deals with a method of solution for *Euler 's stationary system:*

$$\nabla.\rho\mathbf{u} = 0, \quad \nabla.[\rho\mathbf{u} \otimes \mathbf{u}] + \nabla p = \mathbf{0}, \quad \nabla.[\rho\mathbf{u}(\frac{\gamma}{\gamma - 1}\frac{p}{\rho} + \frac{1}{2}u^2)] = 0 \text{ in } \Omega, \qquad (4)$$

where $\mathbf{u.n}$ is given on all of Γ and $\rho, \mathbf{u}$ and p are given on parts of Γ according to the signs of $\mathbf{u.n} \pm (\gamma\frac{p}{\rho})^{1/2}$.

Chapter 3 deals with convection - diffusion phenomena, i.e. equations of the type :

$$\frac{\partial \varphi}{\partial t} + \nabla.\varphi\mathbf{u} - k\Delta\varphi = f \text{ in } \Omega, \tag{5}$$

with φ given on the boundary and at an initial time. The methods of upwinding and artificial viscosity are introduced.

In *chapter 4* we give some finite element methods to solve the generalized Stokes problem :

$$\alpha\mathbf{u} - \Delta\mathbf{u} + \nabla p = \mathbf{f} \quad , \nabla.\mathbf{u} = 0 \text{ in } \Omega, \tag{6}$$

with $\mathbf{u}$ given on the boundary. This problem which is in itself useful for low Reynolds number flows is also utilized as an intermediate step for the resolution of Navier-Stokes equations.

The *chapter 5* shows how one solves the Navier-Stokes equations using the methods given in chapters 3 and 4. Some models of turbulence as well as the methods for solving them are given.

Finally, in *chapter 6* we give a few methods to solve the compressible Euler equations ((7)-(9) with $\mu = \xi = \kappa = 0$) and the Navier-Stokes equations :

$$\frac{\partial \rho}{\partial t} + \nabla.\rho\mathbf{u} = 0 \tag{7}$$

$$\frac{\partial \rho\mathbf{u}}{\partial t} + \nabla.(\rho\mathbf{u} \otimes \mathbf{u}) + \nabla p - \eta\Delta\mathbf{u} - (\frac{\eta}{3} + \xi)\nabla(\nabla.\mathbf{u}) = 0 \tag{8}$$

$$\frac{\partial}{\partial t}[\rho\frac{u^2}{2} + \frac{p}{\gamma - 1}] + \nabla.\{\rho\mathbf{u}[\frac{u^2}{2} + \frac{\gamma}{\gamma - 1}\frac{p}{\rho}]\} \tag{9}$$

$$= \nabla.\{\frac{\kappa}{R}\nabla\frac{p}{\rho} + [\eta(\nabla\mathbf{u} + \nabla\mathbf{u}^T) + (\xi - \frac{2}{3}\eta)\mathbf{I}\nabla.\mathbf{u}]\mathbf{u}\} + \mathbf{f.u}$$

with $\mathbf{u}, p$ given on the boundary, ρ given on part of the boundary and at an initial time.

In this same chapter, we also consider the case of Saint-Venant's shallow water equations because they form an important bidimensional approximation to the incompressible Navier-Stokes equations and they are of the same type as the compressible Navier-Stokes equations.

The book ends with an appendix describing a computer program, written by the author to illustrate the course and used to solve some of the problems studied. Unfortunately, this software makes extensive use of the 'Toolbox' of the Apple Macintosh^{TM} and it is not portable to other machines.

A FEW BOOKS FOR FURTHER READING

[1] A. Baker:*Finite element computational fluid mechanics.* McGraw Hill, 1985.

[2] J.P. Benque, A. Haugel, P.L. Viollet:*Engineering applications of hydraulics II.* Pitman, 1982.

[3] G. Chavent, G. Jaffrey: *Mathematical methods and finite elements for reservoir simulations.* North-Holland, 1986.

[4] T. Chung: *Finite element analysis in fluid mechanics.* McGraw Hill, 1978

[5] C. Cuvelier, A. Seigal, A. van Steenhoven: *Finite element methods and Navier-Stokes equations.* Mathematics and its applications series, D. Reidel Publishing co, 1986.

[6] G. Dhatt, G. Touzot: Une présentation de la méthode des éléments finis. Editions Maloine, 1984.

[7] V. Girault, P.A. Raviart:*Finite element method for Navier-Stokes equations.* Springer Series SCM, **5**, 1985.

[8] R. Glowinski: *Numerical methods for nonlinear variational problems.* Springer Series in comp. physics. 1984.

[9] R. Gruber, J. Rappaz: Finite element methods in linear magnetohydrodynamics. Springer series in comp. physics, 1985.

[10] C. Hirsch: *Numerical computation of internal and external flows.* Wiley, 1988.

[11] T.J.R. Hughes: *The finite element method.* Prentice Hall, 1987.

[12] C. Johnson:*Numerical solution of PDE by the finite element method.* Cambridge university press, 1987.

[13] R. Peyret, T. Taylor: *Computational methods for fluid flows.* Springer series in computational physics. 1985.

[14] F. Thomasset: *Implementation of the finite element method for the Navier-Stokes equations.* Springer Series in comp. physics. 1981.

Chapter 1

Some equations of fluid mechanics

1. ORIENTATION.

The aim of this chapter is to recall the basic equations of fluid mechanics as well as the approximation principles which we propose for their numerical simulations. For more details, see for example Anderson [4], Bachelor [10], Landau-Lifchitz [141], Panton [187].

2. GENERAL EQUATIONS OF NEWTONIAN FLUIDS :

2.1 Equation of conservation of mass .

Let $\rho(x,t)$ be the density of a fluid at a point x at time t ; let $u(x,t)$ be its velocity.

Let Ω be the domain occupied by the fluid and O a regular subdomain of Ω. To conserve the mass, the rate of change of mass of the fluid in O, $\partial(\int_O \rho)/\partial t$, has to be equal to the mass flux across the boundary ∂O of O, $-\int_{\partial O} \rho u.n$, (n denotes the exterior normal at ∂O and the surface element)

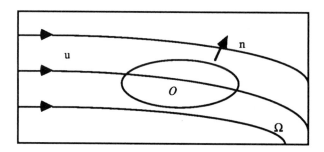

Figure 1.1 : *A volume element of fluid O.*

Using Stokes formula

$$\int_O \nabla.v = \int_{\partial O} v.n \tag{1}$$

and the fact that O is arbitrary, we get immediately the *equation of conservation of mass*

$$\rho_{,t} + \nabla.(\rho u) = 0 \tag{2}$$

2.2 Equation of conservation of momentum

Now let us write Newton's equations for a volume element O of the fluid.

A particle of the fluid at x at an instant t will be at $x + u(x,t)k + o(k)$ at instant $t + k$; its acceleration is therefore

$$\lim_{k \to 0} \frac{1}{k}[u(x + u(x,t)k, t + k) - u(x,t)] = u_{,t} + u_j u_{,j} \tag{3}$$

The external forces on O are :

- The external forces on a unit volume (electromagnetism, Coriolis, gravity...): $\int_O f$
- The pressure forces: $\int_{\partial O} pn$
-The viscous constraints due to deformation of the fluid (σ' is a tensor): .

$$-\int_{\partial O} \sigma'n.$$

Newton's laws are, therefore, written

$$\int_O \rho(u_{,t} + u\nabla u) = \int_O f - \int_{\partial O}(pn - \sigma'n),$$

or, by using Stokes formula

$$\int_O \rho(u_{,t} + u\nabla u) = \int_O (f - \nabla p + \nabla.\sigma'),$$

which gives :

$$\rho(u_{,t} + u\nabla u) + \nabla p - \nabla.\sigma' = f.$$

To proceed further we need a hypothesis to relate the stress tensor σ' with the velocity of the fluid. By definition of the word "fluid", any force, even infinitesimal, applied to O should produce a displacement, thus σ' being the effect due to the viscosity of the fluid, it must depend on the derivatives of the velocity and tend to zero with them.

The hypothesis of *Newtonian flow* is a linear law relating σ' to ∇u , (I being the unit tensor) :

$$\sigma' = \eta(\nabla u + \nabla u^T) + (\xi - \frac{2\eta}{3})I\nabla.u \qquad (4)$$

η and ξ are the first and second viscosities of the fluid.

From (1), (3) and (4) we derive the *equation of conservation of momentum*

$$\rho(u_{,t} + u\nabla u) + \nabla p - \nabla.[\eta(\nabla u + \nabla u^T) + (\xi - \frac{2\eta}{3})I\nabla.u] = f. \qquad (5)$$

It is easy to verify that the equations can also be written as

$$\rho(u_{,t} + u\nabla u) - \eta\Delta u - (\frac{\eta}{3} + \xi)\nabla(\nabla.u) + \nabla p = f \qquad (6)$$

because $\nabla.\nabla u = \Delta u$ and $\nabla.\nabla u^T = \nabla(\nabla.u)$. Taking into account the continuity equation (2), we could also rewrite (6) in a conservative form :

$$\frac{\partial \rho u}{\partial t} + \nabla.(\rho u \otimes u) + \nabla p - \eta\Delta u - (\frac{\eta}{3} + \xi)\nabla(\nabla.u) = f. \qquad (7)$$

Remark 1 :
One can show (see Ciarlet [55], for example) that one cannot put some arbitrary relation between σ' and ∇u or else the equations are no longer invariant under changes of reference frame (translation and rotation), that is to say that they depend on the system of coordinates in which they are written . Thus, at least in 2D, (4) is relatively general and in fact the only relation possible when one assumes that σ' depends on ∇u for an isotropic fluid. Let us give examples of fluids which do not obey the above law :

- Pasty fluids (near to a state of solidification)
- micro-fibers in fluids (some biomechanics fluids with long molecules),

- Mixture of fluid-particles (blood, for example),
- Rarefied gases.

2.3 Equation of conservation of energy and the state equation .

Lastly, an equation called *conservation of energy* can be obtained by writing the total energy of a volume element $O(t)$ moving with the fluid.

We note that the energy E is the sum of the work done by the forces and the amount of heat received. The energy of a volume element O is the sum of potential energy ρe and the kinetic energy $\rho u^2/2$, $\int_O \rho(u^2/2 +e)$. The work done by the forces is : $\int_O u.f + \int_{\partial O} u\sigma'.n - \int_{\partial O} pn.u$.

If there is no *source of heat* (combustion...) then the amount of heat received (lost) is proportional to the gradient flux of the temperature θ : $\int_{\partial O} \kappa \nabla \theta.n$. We therefore get the equation :

$$\frac{d}{dt} \int_{O(t)} [\rho(e + \frac{u^2}{2})] = \int_O \{[\rho(e + \frac{u^2}{2})]_{,t} + \nabla.[u\rho(\frac{u^2}{2} + e)]\}$$

$$= \int_O u.f - \int_{\partial O} [u(\sigma' - p\mathbf{I}) + \kappa \nabla \theta)n$$

With (1) and Stokes' formula, we obtain :

$$\frac{\partial}{\partial t}[\rho(e + \frac{u^2}{2})] + \nabla.(u[\rho(\frac{u^2}{2} + e) + p]) = \nabla.(u\sigma' + \kappa \nabla \theta) + f.u \qquad (8)$$

For an *ideal fluid* , C_v and C_p being the physical constants, we have

$$e = C_v \theta \qquad (9)$$

and *the equation of state*

$$\frac{p}{\rho} = R\theta \qquad (10)$$

where R is an ideal gas constant. With $\gamma = C_p/C_v = R/C_v + 1$ we can write (9) as follows :

$$e = \frac{p}{\rho(\gamma - 1)}. \qquad (11)$$

With (4), (8) becomes :

$$\frac{\partial}{\partial t}[\rho \frac{u^2}{2} + \frac{p}{\gamma - 1}] + \nabla.\{u[\rho \frac{u^2}{2} + \frac{\gamma}{\gamma - 1}p]\} + f.u \qquad (12)$$

$$= \nabla.\{\kappa \nabla \theta + [\eta(\nabla u + \nabla u^T) + (\xi - \frac{2}{3}\eta)\mathbf{I}\nabla.u]u\}$$

This equation can be rewritten as a function of temperature

$$\theta_{,t} + u\nabla\theta + (\gamma - 1)\theta\nabla.u - \frac{\kappa}{\rho C_v}\Delta\theta \tag{13}$$

$$= \frac{1}{\rho C_v}[|\nabla.u|^2(\xi - 2\eta/3) + |\nabla u + \nabla u^T|^2\frac{\eta}{2}].$$

By introducing entropy :

$$s = \frac{R}{\gamma - 1}log\frac{p}{\rho^\gamma} \tag{14}$$

-(12) can be rewritten in the form (Cf. Landau-Lifchitz [141 p. 236]):

$$\rho\theta(\frac{\partial s}{\partial t} + u\nabla s) = \frac{\eta}{2}|\nabla u + \nabla u^T|^2 + (\xi - \frac{2}{3}\eta)|\nabla.u|^2 + \kappa\Delta\theta \tag{15}$$

Remark 2 :
Some values of the physical constants

	$\rho(g/cm^3)$	$\eta(g/cm.s)$	ξ	$\kappa(cm^2/s)$	γ	$R(cm^2/s^2.{}^\circ C)$
air	1.210^{-3}	1.810^{-4}	$\cong 0$	0.2	1.4	2.8710^6
water	1	0.01	$\cong 0$	1.410^{-3}	1	0.2410

The system (2), (5), (10), (12), is called *compressible Navier-Stokes equations*. With suitable boundary and initial conditions, it is complete in the sense that we have 6 unknown scalars (ρ, u, p, θ) and 6 scalar equations ; Matsumura -Nishida[168] and Valli [231] have shown that the system is well posed with regular initial conditions $(\rho^0, u^0, p^0, \theta^0)$ satisfying (10) and regular boundary conditions :
- u, θ given on the boundary, p calculated from (10)
- ρ given on the part of the boundary in which $u.n < 0$.

A number of numerical simulations have been made (mostly using finite differences) but they are extremely onerous and limited to quasi non-stationary laminar flows.

For these reasons and also with a view towards getting partial analytical solutions, the fluid mechanicists have proposed some approximations to the complete system ; it is, in general, these approximated systems that we solve numerically.

3. INVISCID FLOWS.

If we neglect the loss of heat by thermal diffusion ($\kappa = 0$) and the viscous effects ($\eta = \xi = 0$) the simplified equations are known as the *compressible Euler equations*. So equation (15) becomes :

$$\frac{\partial s}{\partial t} + u\nabla s = 0 \qquad (16)$$

hence s is constant on the lines tangent at each point to u (stream lines). In fact a stream line is a solution of the equation :

$$x'(\tau) = u(x(\tau), \tau) \qquad (17)$$

and so

$$\frac{d}{dt} s(x(t), t) = \frac{\partial s}{\partial x_i}\frac{\partial x_i}{\partial t} + \frac{\partial s}{\partial t} = s_{,t} + u\nabla s = 0. \qquad (18)$$

If s is constant and equal to s^0 constant at time 0 and if s is also equal to s^0 on the part of Γ where $u.n < 0$, then from (17) we see that (18) has an analytical solution $s = s^0$. Note however that (18) has no meaning if u has a shock. In any case, experiments show that s decreases through shocks.

Finally there remains a system of two equations with two unknowns :

$$\rho_{,t} + \nabla.(\rho u) = 0 \qquad (19)$$

$$\rho(u_{,t} + u\nabla u) + \nabla p = f \qquad (20)$$

where

$$p = C\rho^\gamma \quad (C = e^{s^0\frac{\gamma-1}{R}}). \qquad (21)$$

The problems (2)(5)(12) and (19)-(21) will be studied in chapter 6 but an algorithm for finding the stationary solutions of (19)-(21) will also be given in Chapter 2.

4.   INCOMPRESSIBLE OR WEAKLY COMPRESSIBLE FLOWS.

In the case when ρ is practically constant (water for example or air with low velocity) we can neglect its derivatives. Then (2), (6) become the incompressible *Navier Stokes equations* :

$$\nabla.u = 0 \qquad (22)$$

$$u_{,t} + u\nabla u + \nabla p - \nu\Delta u = f \qquad (23)$$

where $\nu = \eta/\rho$ is the *reduced viscosity* of the fluid and p/ρ and f/ρ have been replaced by p and f.

The same approximation can be applied to equations (19)-(20) and we obtain (22)-(23) with $\nu = 0$, known as the incompressible *Euler equations.*

$$\nabla.u = 0 \qquad (24)$$

$$u_{,t} + u\nabla u + \nabla p = f \tag{25}$$

The Navier-Stokes and Euler's equations will be studied in chapter 5.

Remark 3.

With (22)(23) or (24)(25) one can no longer use (21) because the system becomes over determined. This ambiguity can be explained by studying (19)(21)when the velocity of sound $(dp/d\rho)^{1/2}$ tends to infinity. One can show (Klainerman -Majda [165]) that if C tends to infinity in (21) then p goes to infinity but ∇p remains bounded. Thus the pressure variation continues to have a physical interpretation in (22)(23) or (24)(25) but p no longer represents the physical pressure.

An equation for the temperature θ can be obtained from (13) by assuming ρ constant; if $f = 0$ we have :

$$\frac{\partial \theta}{\partial t} + u\nabla\theta - \frac{\kappa}{\rho C_v}\Delta\theta = \frac{\nu}{2C_v}|\nabla u + \nabla u^T|^2 \tag{26}$$

An intermediate approximation, called *weakly compressible* , in between the complete system (2), (5), (12) (compressible Navier-Stokes) and incompressible Navier-Stokes (22)-(23), is interesting because it exhibits the *hyperbolic character* of the underlying acoustics in (2), (5), (12) and (19)(21).

Assume that $\xi = \eta = \kappa = 0$, which implies s constant and so $p\rho^{-\gamma}$ constant; under the assumption that ρ oscillates around a value ρ^0 one gets :

$$\rho_{,t} + u\nabla\rho + \rho^0\nabla.u = 0$$
$$u_{,t} + u\nabla u + C\rho^{0\gamma-1}\nabla\rho = 0. \tag{27}$$

If also u and $\rho - \rho^0$ are small, then (27) is close to

$$\rho_{,t} + \rho^0\nabla.u = 0; \quad u_{,t} + C\rho^{0\gamma-1}\nabla\rho = 0 \tag{28}$$

which gives

$$\rho_{,tt} - C\rho^{0\gamma}\Delta\rho = 0. \tag{29}$$

Finally, if one is interested in *thermal convection* problems in the fluid (which happens as water is heated in a pan, for example) one could assume that ρ is quasi.constant in (7) and could neglect all variations of ρ except of f/ρ in (5) where f is gravity. Then one obtains the Rayleigh-Benard equations :

$$\nabla.u = 0$$
$$u_{,t} + u\nabla u + \nabla p - \nu\Delta u = -g\theta e_3 \tag{30}$$
$$\theta_{,t} + u\nabla\theta - \kappa'\Delta\theta = \frac{\nu}{2C_v}|\nabla u + \nabla u^T|^2$$

where e_3 is the unit vector in the vertical direction.

5. IRROTATIONAL FLOWS :

We could try to determine if, for suitable boundary conditions, there exist solutions of equations satisfying

$$\nabla \times u = 0; \tag{31}$$

these solutions are called *irrotational*.

But as (22) implies the existence of φ (x,t) such that

$$u = \nabla\varphi \tag{32}$$

these solutions are also called *potential*.

Using the identities :

$$\Delta u = -\nabla \times \nabla \times u + \nabla(\nabla.u) \tag{33}$$

$$u\nabla u = -u \times (\nabla \times u) + \nabla(\frac{u^2}{2}) \tag{34}$$

we say that (24)-(25) have such solutions for $f = 0$ if φ is a solution of the *Laplace equation* :

$$\Delta\varphi = 0, \tag{35}$$

with (32) and

$$p = k - \frac{1}{2}|\nabla\varphi|^2. \tag{36}$$

This type of flow is the simplest of all. It will be studied in chapter 2.

In the same way, with (32), (27) implies

$$\rho_{,t} + \nabla\varphi\nabla\rho + \rho^0\Delta\varphi = 0 \tag{37}$$

$$\nabla(\varphi_{,t} + \frac{1}{2}|\nabla\varphi|^2 + \gamma C\rho^{0\gamma-1}p) = 0 \tag{38}$$

If we neglect the convection term $\nabla\varphi\nabla\rho$ this system simplifies to a nonlinear *wave equation* :

$$\varphi_{,tt} - c\Delta\varphi + \frac{1}{2}|\nabla\varphi|^2_{,t} = d(t) \tag{39}$$

where c $= \gamma C\rho^{0\gamma}$ is related to the velocity of the sound in the fluid.

Finally, we show that there exist stationary potential solutions of (19), (20), (21) with $f = 0$. Using (34), (19) could be rewritten as follows :

$$-\rho u \times \nabla \times u + \rho \nabla \frac{u^2}{2} + \nabla p = 0. \tag{40}$$

Taking the scalar product with u, we obtain

$$u.[\rho \nabla \frac{u^2}{2} + \nabla p] = 0 \tag{41}$$

Also, on taking into account (21)

$$u.(\rho \nabla \frac{u^2}{2} + \rho^{\gamma-1} C\gamma \nabla \rho) = 0 \tag{42}$$

Or

$$u\rho.\nabla(\frac{u^2}{2} + C\frac{\gamma}{\gamma - 1}\rho^{\gamma-1}) = 0 \tag{43}$$

So, the quantity between the parenthesis is constant along the stream lines, that is we have

$$\rho = \rho^0(k - \frac{u^2}{2})^{\frac{1}{\gamma-1}} \tag{44}$$

Indeed a system like $u\nabla\xi = 0$ is integrated exactly like (16) and gives ξ constant if it is constant on the part of the boundary where $u.n < 0$. Thus if ρ^0 and k are constant at the entrance boundary of Ω ($u.n¡0$), if all the streamlines intersect the entrance boundary of Ω and if u is non zero, then (44) holds, then (40) implies that $\nabla \times u$ is parallel to u and, at least in 2 dimensions , this implies that u derives from a potential (i.e. (31)). From (2) and (44) we deduce the *transonic potential flow equation.*

$$\nabla.[(k - |\nabla\varphi|^2)^{\frac{1}{\gamma-1}}\nabla\varphi] = 0 \tag{45}$$

This problem will be studied in chapter 2.

Remark 4:
Equation (35) is consistent with (45) since we can return to it by assuming ρ constant (Cf (44)).

6. THE STOKES PROBLEM.

Let us come back to the system (19)-(21) and let us rewrite it in non dimensional form.

Let U be a characteristic velocity of the flow under study (for example one of the non homogeneous boundary conditions).

Let L be the characteristic length (for example the diameter of Ω) and T a characteristic time (which is a priori equal to L/U).Let us put

$$u' = \frac{u}{U}; \quad x' = \frac{x}{L}; \quad t' = \frac{t}{T} \tag{46}$$

Then (15) and (16) can be rewritten as

$$\nabla_{x'}.u' = 0 \tag{47}$$

$$(\frac{L}{TU})u'_{,t'} + u'\nabla_{x'}u' + U^{-2}\nabla_{x'}p - (\frac{\nu}{LU})\Delta_{x'}u' = f\frac{L}{U^2} \tag{48}$$

So, if we put $T = L/U$, p' $= p/U^2$

$$\nu' = \frac{\nu}{LU} \tag{49}$$

then (38)-(39) is the same as (15)-(16) with "prime" variables.The inverse of ν' is called *Reynolds number*. Let us give some examples :

$(unitsMKS)$	U	L	ν	$Re = \frac{UL}{\nu}$
$micro-organism$	10^{-4}	10^{-4}	10^{-6}	10^{-2}
glider	1	1	0.15×10^{-4}	7×10^4
sailboat(shell)	0,1	1	10^{-6}	10^5
car	3	3	0.15×10^{-4}	6×10^5
airplane	30	10	0.15×10^{-4}	2×10^7
tanker	1	200	10^{-6}	2×10^8

When $\nu' \gg 1$, $\nu'\Delta u'$ dominates $u'\nabla u'$ and $u'_{t'}$ in (48); it becomes the *Stokes problem* :

$$-\nu\Delta u + \nabla p = f \tag{50}$$

$$\nabla.u = 0 \tag{51}$$

Experience shows that (50)-(51) is indeed an excellent approximation of (15)-(16) for "*low Reynolds number flows*" in dimension three ; in dimension two the Stokes problem might not have solution in an unbounded domain. We recall in this context that a two dimensional flow is not an infinitely thin flow, but a flow invariant under translation in a spatial direction and whose velocity is perpendicular to the invariant direction. .

The Stokes problem will be studied in chapter 4.

7. CHOICE OF EQUATIONS.

Given a fluid system to simulate numerically, what equations need to be chosen? Unfortunately, there is no automatic rule ; one seeks, in general the simplest system compatible with the phenomenon.

Let us take the case of an aircraft wing. The first thing that an engineer wishes to know is evidently the resultant of the fluid forces on the wing ; that is the lift and the drag. The lift being somewhat unrelated to the viscosity we can integrate the transonic equation to calculate it (cf Chapter 2). But if there is a lot of vorticity then we have to take the Euler equation, compressible if the velocity if large, incompressible if not. On the other hand, there is no chance to calculate the drag with Euler's equations; we have to integrate the Navier-Stokes equations (in the boundary layer or in the whole domain). The problem becomes more complicated if the flow is "detached"; though the lift is not a viscous phenomenon, the generation of eddies near the trailing edge is fundamentally a phenomenon related to the viscosity. The oscillation of the lift around its average value cannot be approximated correctly unless the Navier-Stokes equations are solved.

To conclude, the choice of a system to solve is beyond the scope of this book and this chapter gives only some examples. A thorough discussion with a physicist or with an engineer is therefore indispensable before embarking on numerical simulations which are often long and expensive.

8 CONCLUSION:

The compressible Navier Stokes equations describe a large number of phenomena which from the mathematical point of view belong to 3 categories of PDE's :

Elliptic: Stokes equations, incompressible potential flow .

Parabolic : Temperature propagation, convection-diffusion. Incompressible Navier-Stokes equation.

Hyperbolic: Euler's equations, wave equations of the acoustics of a fluid, convection.

We state also, that there does not exist a universal scheme to approximate numerically these three types of equations, which explains why each subsystem enumerated above and its numerical integration is treated in a separate chapter of this book.

Chapter 2

Irrotational and weakly irrotational flows

1. ORIENTATION

We have seen in chapter 1 that it is possible (with suitable boundary conditions) to find irrotational solutions ($\nabla \times u = 0$) to the general fluid mechanics equations under the following hypotheses :
- inviscid flow
- no vorticity generated by discontinuities (shocks) or by the boundaries.

In this chapter, we will study finite element approximations of those equations. The first part deals with the Neumann problem, where we recall also the finite element method. In the second part, we deal with subsonic compressible flows, which will be solved by an optimization method. We then extend the method to transonic flows. Finally in the third and fourth parts we study the resolution of the problem in terms of stream function and a rotational correction method based on Helmoltz decomposition of vector fields.

2.INCOMPRESSIBLE POTENTIAL FLOWS

2.1 Generalities.

If the density ρ is constant, the viscosity and thermal diffusion are negligible and if the flow does not depend on time then the velocity $u(x)$ and pressure $p(x)$ are given for all points $x \in \Omega$ of the fluid by

$$\nabla . u = 0 \qquad (1)$$

$$\nabla \times u = 0 \tag{2}$$

$$p = k - \frac{1}{2}u^2 \tag{3}$$

where k is a constant if the flow is uniform at infinity.

On the boundary $\Gamma = \partial\Omega$ in general the normal component of the velocity, $u.n$, is given by :

$$u.n = g \quad on \quad \Gamma \tag{4}$$

We remark that (2) implies the existence of a potential φ such that :

$$u = \nabla\varphi \tag{5}$$

So the complete system is simply rewritten as a Neumann problem.

$$\Delta\varphi = 0 \quad in \quad \Omega \tag{6}$$

$$\frac{\partial\varphi}{\partial n} = g \quad on \quad \Gamma \tag{7}$$

Example 1.
Calculation of low velocity laminar flow through a nozzle.

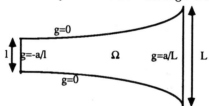

Figure 2.1 . *A divergent nozzle*

Example 2
Calculation of the flow around a symmetric wing (the non-symmetric i.e. lifting case will be studied later).

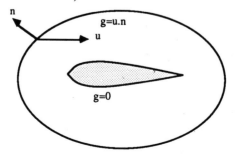

Figure 2.2 *A symmetric airfoil*

Remark

We could have used (1) to say that there exists ψ such that

$$u = \nabla \times \psi \tag{9}$$

Then, the complete system becomes

$$\nabla \times \nabla \times \psi = 0 \quad in \quad \Omega \tag{10}$$

$$\nabla \times \psi.n = g \quad on \quad \Gamma \tag{11}$$

If the flow is bidimensional (invariant by translation in one direction) then (10)-(11) become :

$$\Delta \psi = 0 \tag{12}$$

$$\psi(x(s)) = \int^s g(x(\gamma)d\gamma \quad \forall x(s) \in \Gamma \tag{13}$$

We study this method at the end of the chapter. Note that in 3 dimensions, the operator in (10)-(11) is not strongly elliptic and that ψ is a vector though φ is a scalar.

2.2. Variational formulation and discretisation of (6)-(7)

Proposition 1

Let $\chi \in L^2(\Omega)$ with non zero average on Ω. If g is regular, $(g \in H^{1/2}(\Gamma))$ and
if

$$\int_\Gamma g = 0 \tag{14}$$

then problem (6)-(7) is equivalent to the following variational problem :

$$\int_\Omega \nabla\varphi\nabla w = \int_\Gamma gw \quad \forall w \in W \tag{15}$$

$$\varphi \in W = \{w \in H^1(\Omega); \ \int_\Omega \chi w = 0\} \tag{16}$$

Proof :

We see, in multiplying (6) by w, and integrating on Ω with the help of Green's formula that :

$$\int_\Omega -(\Delta\varphi)w + \int_\Gamma \frac{\partial\varphi}{\partial n}w = \int_\Omega \nabla\varphi\nabla w. \tag{17}$$

If we use (6) and (7) in (17) we get

$$\int_{\Omega} \nabla\varphi\nabla w = \int_{\Gamma} gw \quad \forall w \in H^1(\Omega) \tag{18}$$

Putting $w = 1$ we get (14). As (18) doesn't change if we replace φ by $\varphi +$ constant and w by $w + constant$, we can restrict ourselves to W. The converse can be proved in the same way ; from (15) with the assumption that φ is regular (in $H^2(\Omega)$) so as to use (17), we get :

$$\int_{\Omega} -(\Delta\varphi)w + \int_{\Gamma} (\frac{\partial\varphi}{\partial n} - g)w = 0 \quad \forall w \in W \tag{19}$$

By taking w to be zero on the boundary, we deduce that :

$$\int_{\Omega} -(\Delta\varphi)w = 0 \quad \forall w \text{ such that } \int_{\Omega} \chi w = 0,$$

and so, from the theory of Lagrange multipliers, there exists a constant λ such that

$$-\Delta\varphi = \lambda\chi \quad in \quad \Omega. \tag{20}$$

Note also that (19) implies

$$\frac{\partial\varphi}{\partial n} = g \quad on \quad \Gamma \tag{21}$$

Finally, (17) with $w = 1$ (20), (21) and (14) imply that the constant in (20) is zero.

Remark
W given by (16) is equivalent to $H^1(\Omega)/R$. One could have set up the problem in $H^1(\Omega)/R$ at the start but this presentation allows a natural construction for the discretisation of W. Besides, it is interesting to note that if (14) is not satisfied, φ is the solution of (20) instead of (6).

Proposition 2
If Ω is a bounded domain with Lipschitz boundary and if g is in $H^{1/2}(\Gamma)$ then (15) has a unique solution.

Proof :
In W and with bounded Ω, Poincaré's inequality is satisfied:

$$\int_{\Omega} \varphi^2 \leq C \int_{\Omega} |\nabla\varphi|^2. \tag{22}$$

So the bilinear form associated with (15) is $W - elliptic$:

$$\int_{\Omega} |\nabla\varphi|^2 \geq \frac{1}{2} \min\{C^{-1}, 1\} \|\varphi\|_1^2 \tag{23}$$

As W is a closed subspace of $H^1(\Omega)$ the linear map

$$w \to \int_\Gamma g w d\gamma$$

is continuous. The result follows from Lax-Milgram theorem (cf Ciarlet [55] Strang-Fix [223], Lions-Magenes [155]).

Proposition 3

Let $\{W_h\}$ be a sequence of internal approximations $(W_h \subset W)$ of W, such that

$$\forall w \in W \ \exists \{w_h\}_h, \quad w_h \in W_h \text{ such that } \|w_h - u\|_1 \to 0 \text{ when } h \to 0 \quad (24)$$

Then the solution $\varphi_h \in W_h$ of

$$\int_\Omega \nabla \varphi_h \nabla w_h = \int_\Gamma g w_h \quad \forall w_h \in W_h \quad (25)$$

converges strongly in $H^1(\Omega)$ to φ, the solution of (15)-(16).

Proof:
Using (15) with w_h and subtracting from (24), we obtain

$$\int_\Omega \nabla(\varphi_h - \varphi)\nabla w_h = 0 \quad \forall w_h \in W_h \quad (26)$$

which shows that φ_h is a solution of the problem

$$\min_{w_h \in W_h} \{\int_\Omega |\nabla(w_h - \varphi)|^2\} \quad (27)$$

and so if ψ_h is an approximation of φ in the sense of (24) we have

$$\int_\Omega |\nabla(\varphi_h - \varphi)|^2 \leq \int_\Omega |\nabla(\psi_h - \varphi)|^2 \quad (28)$$

and hence the proof (Cf. (22)).

Proposition 4

If there exist α and $C(\Omega)$ such that in addition to (24) we have

$$\|w_h - u\|_1 \leq C h^\alpha |u|_{\alpha+1} \quad (29)$$

$$\text{then } \|\varphi_h - \varphi\|_1 \leq C' h^\alpha \|\varphi\|_{\alpha+1} \quad (30)$$

Remark
If Ω is convex, we have also

$$|\varphi_h - \varphi|_0 \leq C'' h^{\alpha+1} \|\varphi\|_{\alpha+1}. \quad (31)$$

Proof :

The error estimate (30) can easily be deduced from (28) (29) and (22). To prove (31) one has to use the duality argument of Aubin-Nitsche (see Ciarlet [56])

Lagrangian triangular finite elements (also called P1 conforming):
Ω is divided into triangles (tetrahedra in 3D) $\{T_k\}_{1...K}$ such that

- $T_k \cap T_l = \emptyset$, or 1 vertex, or 1 whole side (resp. side or face) when $k \neq l$
- The vertices of the boundary of $\cup T_k$ are on Γ
- The singular points of Γ (corners) are on the boundary Γ_h of $\cup T_k$.

We note that $\Omega_h = \cup T_k$, $\Gamma_h = \partial \cup T_k$, $\{q^i\}_1^N$ are the vertices of the triangles, and h is the longest side of a triangle :

$$h = \max_{\{i,j,k:\, q^i, q^j \in T_k\}} |q^i - q^j| \qquad (32)$$

$$H_h = \{u_h \in C^0(\Omega_h) : u_h|_{T_k} \in P_\alpha\} \qquad (33)$$

$$W_h = \{u_h \in H_h : \int_{\Omega_h} \chi u_h = 0\} \qquad (34)$$

where P_α denotes the space of polynomials in n variables of degree less than or equal to α ($\Omega \subset R^n$) and C^0 the space of continuous functions.

If Ω is a polygon then $W_h \subset W$ and proposition 3 applies. If moreover, we assume that all the angles of the triangles are bounded by $\theta_1 > 0$ and $\theta_2 < \pi$ when $h \to 0$ then the proposition 4 can be applied.

With (33) W_h is of finite dimension, say M-1. Let $\{u'^i\}_{1...M-1}$ be a basis of W_h; if we write φ_h in this basis, we have

$$\varphi_h(x) = \sum_{1}^{M-1} u'^i \varphi'_i(x) \qquad (35)$$

and if we replace u'_h by u'^i in (25) we get a $(M-1) \times (M-1)$ linear system.

$$\mathbf{A\Phi} = G \qquad (36)$$

$$\text{with } \Phi = \{\varphi_1...\varphi_{M-1}\}, \quad A_{ij} = \int_\Omega \nabla u'^i \nabla u'^j, \quad G_j = \int_\Gamma gu'^j \qquad (37)$$

Example 1

With $\alpha = 1$ a basis of W_h can easily be constructed. Let $\{\lambda^i_j(x)\}_{i=1...n+1}$ denote the barycentric coordinates of x in T_k with respect to its $n+1$ vertices then the number of basis functions is $N - 1$ (i.e. $M = N$) that is one less than the total number of vertices $\{q^i\}_{i=1...N}$. Let u^i be the canonical basis function of Lagrangian elements of order 1:

$$w_k^i(x) = \lambda_k^i(x) \text{ if } k \text{ is such that } x \in T_k, \quad (q^i \in T_k) \tag{38}$$

$$= 0 \quad otherwise$$

then

$$w^i(x) = w_k^i(x) - \frac{\int_\chi w_k^i(x)}{\int_\chi \chi} \qquad i = 1..N - 1 \tag{39}$$

is a basis of W_h.

Remark :

As w^i differs from w_k^i only by a constant, thanks to (14), one can use either w^i or w^{i} in (37). It makes no difference. The effect of this construction is simply to pull out a function w^i to construct a basis. This procedure has the inconvenience of distinguishing a vertex from others (since we have pulled the w^i associated to it) and sometimes introduces a local numerical error around the pulled vertex because the linear system is poorly conditioned there. We will see below that the conjugate gradient method avoids this inconvenience.

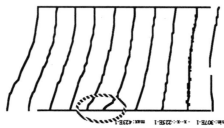

Figure 2.3 : *Example of numerical singularity obtained by pulling a basis function ; we remark in the top middle a distortion of isovalue lines .*

Example 2

With $\alpha = 2$ the number of basis functions is equal to the number of vertices q^i plus the midpoints of the sides (edges) minus one. We can also construct it by (39) but with λ_k^i replaced by $\lambda_k^i(2\lambda_k^i - 1)$ for the basis functions associated to vertices and by $4\lambda_k^i\lambda_k^j$ for the basis functions associated to the midpoints of the sides (edges).

2.3. Resolution of the linear system by the conjugate gradient method.

The linear system (36) obtained by the finite element method has a peculiar structure which we must exploit to optimize the resources of the computer. The conjugate gradient method (cf. Polak [193], Lascaux-Théodor [142], Luenberger [162] for example) makes use of the sparse structure of the linear system to

speed the solution when properly preconditioned. This algorithm applies also to positive semi-definite linear systems; thus we can construct the system (36), (37) with *all* the basis functions u^i. The solution obtained is more regular (cf. figures 3, 4).

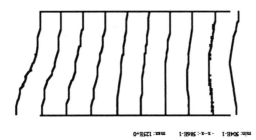

min: 504E-1 · -x-x-: 586E-1   max: 125E+0

Figure 2.4: *Result obtained by the conjugate gradient method without removing the basis function.*

Let us recall the conjugate gradient method briefly.

Notation:
Let $x \in R^N$ be an unknown such that $Ax = b$, where $A \in R^{N \times N}$, $A^T = A$, $b \in R^N$.

Let C be a positive definite matrix and $<\ ,\ >_C$ the scalar product associated with C:

$$<a,b>_C = a^T C b; \quad \|a\|_C = (a^T C a)^{\frac{1}{2}}$$
(40)

Algorithm 1
0. Initialization:
Choose $C \in R^{N \times N}$ positive definite (preconditioning matrix), ϵ small positive, x^0 such that $Cx^0 \in Im A$ (0 for example), and set $g^0 = h^0 = -C^{-1}(Ax^0 - b)$, $n = 0$.

1 Calculate :

$$\rho^n = \frac{<g^n, h^n>_C}{h^{nT} A h^n}$$
(41)

$$x^{n+1} = x^n + \rho^n h^n$$
(42)

$$g^{n+1} = g^n - \rho^n C^{-1} A h^n$$
(43)

$$\gamma^n = \frac{\|g^{n+1}\|_C^2}{\|g^n\|_C^2}$$
(44)

$$h^{n+1} = g^{n+1} + \gamma^n h^n$$
(45)

2 If $\|g^{n+1}\|_C < \epsilon$, stop else increment n by 1 and go to 1.

Note that C is never inverted and that $y = C^{-1} z$ *is a short hand notation for "solve Cy = z"*

Remark 1

This algorithm can be viewed as a particular case of a more general algorithm used to find the minimum of a function (see below). Here it is applied to the computation of the minimum of $E(x) = x^T A x / 2 - b^T x$. One can verify that

i) ρ^n given by (1) is also the minimum of $E(x^n + \rho h^n)$

ii) $-g^{n+1}$ given by (43) is also the gradient of E with respect to the scalar product associated with C, that is :

$$g^{n+1} = -C^{-1}(Ax^{n+1} - b) \tag{46}$$

Proof :

$$x^{n+1} = x^n + \rho^n h^n \iff Ax^{n+1} = Ax^n + \rho^n Ah^n$$

$$\iff C^{-1}(Ax^{n+1} - b) = C^{-1}(Ax^n - b) + \rho^n C^{-1} Ah^n$$

Remark 2

The only divisions in the algorithm are by $h^{nT} Ah^n$ and $\|g^n\|_C$. The first one could be zero if the kernel of A is non empty whereas the second is non zero by construction; so to prove that the algorithm is applicable even if $det(A) = 0$ we have to prove that $h^{nT} Ah^n$ is never zero. But before that let us show convergence.

Lemma

$$h^{jT} Ah^k = 0, \quad \forall j < k \tag{47}$$

$$< g^k, g^j >_C = 0, \quad \forall j < k \tag{48}$$

$$< g^k, h^j >_C = 0, \quad \forall j < k \tag{49}$$

Proof :

Let us proceed by the method of induction. Assuming that the property is true for $j < k \le n$, let us prove that (47)(48)(49) are true for all $j < k \le n+1$

i) Multiplying (43) by h^j and using (47) and (49):

$$< g^{n+1}, h^j >_C = < g^n, h^j >_C - \rho^n < C^{-1} Ah^n, h^j >_C = 0 - h^{jT} Ah^n = 0, \tag{50}$$

if $j < n$; If $j = n$ then it is zero by (41).

ii) Using (45):

$$< g^{n+1}, g^j >_C = < g^{n+1}, h^j - \gamma^j h^{j-1} >_C = 0 \qquad (51)$$

by i) above.

iii) Finally, again from (45) we have, if $j < n$:

$$h^{n+1T} Ah^j = (g^{n+1} + \gamma^n h^n)^T Ah^j = g^{n+1T} Ah^j + 0 = g^{n+1T} C \frac{(g^{j+1} - g^j)}{\rho^j} = 0 \qquad (52)$$

where we have used (47) to get the second equality and (43) for the last one.

If $j = n$, we have to use the definition of γ^n. Let us show that γ^n is also equal to $-g^{n+1T} Ah^n / h^{nT} Ah^n$. We have :

$$\frac{g^{n+1T} Ah^n}{h^{nT} Ah^n} = \frac{< g^{n+1} - g^n, g^{n+1} >_C}{< g^{n+1} - g^n, h^n >_C} = \frac{-\|g^{n+1}\|_C^2}{< g^n, h^n >_C} \qquad (53)$$

We have used (43) for the first equation and (48) and (49) for the second.

We leave to the reader the task of showing that (47),(49) are true for $k = 1$. We end the proof by showing that $h^{nT} Ah^n$ is never zero.

Indeed if n is the first time it is zero then we have:

$$0 = h^{nT} Ah^n, \quad h^n \in ImA \quad \Rightarrow \quad h^n = 0 \quad \Rightarrow \quad g^n = -\gamma^{n-1} h^{n-1}$$

$$= g^{n-1} - \rho^{n-1} C^{-1} Ah^{n-1}$$

so by (41)

$$-\gamma^{n-1} |h^n|^2 = < g^{n-1} h^{n-1} > -\rho^{n-1} h^{n-1T} Ah^{n-1} = 0$$

but by hypothesis h^{n-1} is not zero.

Corollary
The algorithm converges in N iterations at the most.

Proof :
Since the g^n are orthogonal there cannot be more than N of them non zero. So at iteration $n = N$, if not before, the algorithm produces $g^n = 0$ that is $C^{-1}(Ax^n - b) = 0$.

Choice of C
However it is out of question to do N iterations because N is a very big number. One can prove the following result :

Proposition 5:
If x^* is the solution and x^k is the computed solution at the kth iteration, then

$$< x^k - x^*, A(x^k - x^*) >_C \leq 4(\frac{\mu_A^C - 1}{\mu_A^C + 1})^{2k} < x^0 - x^*, A(x^0 - x^*) >_C$$

where μ_A^C is the condition number of A (ratio between the largest and smallest eigenvalues of $Az = \lambda Cz$) in the metric introduced by C.

Proof : see Lascaux-Theodor [9], for example.

One can also prove superconvergence results, that is the sequence $\{x^k\}$ converges faster than all the geometric progressions (faster than r^k for all r) but this result assumes that the number of iterations is large with respect to N.

We can easily verify that if $x^0 = 0$ and $C = A$ we get the solution in one iteration. So this is an indication that one should choose C 'near' to A. When we do not have any information about A, experience shows that the following choices are good in increasing order of complexity and performance:

$$C_{ij} = A_{ii}\delta_{ij} \tag{54}$$

$$C_{ij} = A_{ij}, \quad \forall i, j \; |i - j| < 2 \tag{55}$$

$$C_{ij} = 0, \quad otherwise.$$

$$C = \text{incompletetely factorized matrix of } A. \tag{56}$$

We recall the principle of incomplete factorization (Meijerink-VanderVorst [172], Glowinski et al [97]):
We construct the Choleski factorization L' of A ($= L'L'^T$) and put

$$L_{ij} = 0 \quad if \quad A_{ij} = 0, \quad L_{ij} = L'_{ij} \quad otherwise.$$

One can also construct directly L instead of L' by putting to 0 all the elements of C' which correspond to a zero element of A, *during* the factorization (cf [10]) but then the final matrix may not be positive definite.

Proposition 6
If A is positive semi definite, b is in the image of A, and Cx^0 is in the image of A, then the conjugate gradient algorithm converges towards the unique solution x' of the linear system $Ax = b$; which verifies $Cx' \in Im$ A.

Proof :
From (42) and (45) we see that

$$Cx^n, Cg^n, Ch^n \in ImA \quad \Rightarrow \quad Cx^{n+1}, Cg^{n+1}, Ch^{n+1} \in ImA$$

The property is thus proved by induction.

2.4 Computation of nozzles

If Ω is a nozzle we take, in general, ϕ zero on the walls of the nozzle and $g = u_\infty.n$, u_∞ constant at the entrance and exit of the nozzle.

The engineer is interested in the pressure on the wall and the velocity field. One sometime solves the same problem with different boundary conditions at the entrance Γ_1 and at the exit Γ_2 of the nozzle :

$$-\Delta\varphi = 0 \quad in \quad \Omega, \quad \varphi|_{\Gamma_1} = 0, \quad \varphi|_{\Gamma_2} = constant, \quad \frac{\partial\varphi}{\partial n}|_{\partial\Omega - \Gamma_1 \cup \Gamma_2} = 0. \quad (57)$$

The same method applies.

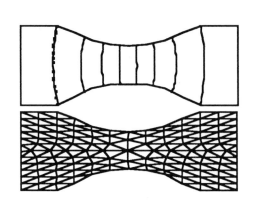

min: 000000   · x–x–: 111E-3   max: 100E-2

Figure 2.5 : *Computation of flow in a nozzle . The triangulation and the level lines of the potential are shown.*

2.5. Computation of the lift of a wing profile

The flow around a wing profile S corresponds, in principle, to flow in an unbounded exterior domain, but we approximate infinity numerically by a boundary Γ_∞ at a finite distance ; so Ω is a two dimensional domain with boundary $\Gamma = \Gamma_\infty \cup S$.

One often takes u_∞ constant and

$$g|_{\Gamma_\infty} = u_\infty.n, \quad g|_S = 0 \tag{58}$$

Unfortunately, the numerical results show that with these boundary conditions, the flow generally goes around the trailing edge P. As P is a singular point of Γ, $|\nabla\varphi(x)|$ tends to infinity when $x \to P$ and the viscosity effects (η and ζ) are no longer negligible in the neighborhood of P (see figure 6). The modeling of the flow by (1)-(2) is not valid and (2) has to be replaced by (ω constant) :

$$\nabla \times u = \omega \delta_\Sigma$$

where δ_Σ is the Dirac function on the stream line Σ which passes through P.

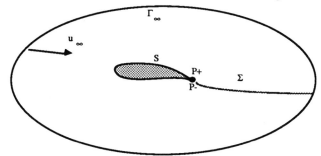

Figure 2.6 *Flow around an airfoil; Σ is the wake.*

One takes then $\Omega - \Sigma$ as the domain of computation. So we have to add a boundary condition on Σ. Since Σ is a stream line :

$$\frac{\partial \varphi}{\partial n}\big|_\Sigma = 0. \qquad (59)$$

But as u is continuous along Σ we also have :

$$\frac{\partial \varphi}{\partial \sigma}\big|_{\Sigma+} = \frac{\partial \varphi}{\partial \sigma}\big|_{\Sigma-}.$$

by integrating this equation on Σ we obtain :

$$\frac{\partial}{\partial \sigma}(\varphi|_{\Sigma+} - \varphi|_{\Sigma-}) = 0 \text{ i.e. for some constant } \beta \ \varphi|_{\Sigma+} - \varphi|_{\Sigma-} = \beta \qquad (60)$$

where β is a constant which is to be determined with the help of (59) written on P, or better by :

$$|\nabla\varphi(P^+)|^2 = |\nabla\varphi(P^-)|^2 \qquad (61)$$

which comes out to be the same but is interpreted as a continuity condition on the pressure.

It can be proved using conformal mappings that with (60)-(61) the solution φ does not depend on the position of Σ (which is not known a priori) but with (59) this is not the case: the solution depends on the position of Σ. So we solve

$$-\Delta\varphi = 0 \quad in \quad \Omega - \Sigma \qquad (62)$$

$$\varphi|_{\Sigma+} - \varphi|_{\Sigma-} = \beta \qquad (63)$$

$$|\nabla\varphi(P^+)|^2 = |\nabla\varphi(P^-)|^2 \qquad (64)$$

$$\frac{\partial \varphi}{\partial n}|_{\partial \Omega} = g; \tag{65}$$

equation (64) which is called the *Joukowski condition* . It deals with the continuity of the velocity and also of the *pressure* .

To solve (62)-(65) a simple method is to note that the solution of (62)-(63)-(65) is linear in β :

$$\varphi(x) = \varphi^0(x) + \beta(\varphi^1(x) - \varphi^0(x)) \tag{66}$$

- φ^0 is the solution of (62),(65) and (53) with $\beta = 0$, i.e.

$$\Delta \varphi^0 = 0, \quad \frac{\partial \varphi}{\partial n}|_\Gamma = g, \quad \varphi \text{ continuous across } \Sigma$$

- $\psi = \varphi^1 - \varphi^0$ is the solution of (62),(65) with $g = 0$ and (63) with $\beta = 1$, i.e.

$$\Delta \psi = 0, \quad \frac{\partial \psi}{\partial n}|_\Gamma = 0, \quad \psi|_{\Sigma+} - \psi|_{\Sigma-} = 1, \quad \nabla \psi|_{\Sigma+} = \nabla \psi|_{\Sigma-}$$

The variational formulation of this second problem is: find $\varphi \in W_1^p$ such that

$$\int_\Omega \nabla \psi \nabla w = 0 \quad \forall w \in W_0^p;$$

$$W_\beta^p = \{\psi \in H^1(\Omega): \quad \psi|_{\Sigma+} - \psi|_{\Sigma-} = \beta, \quad \nabla \psi|_{\Sigma+} = \nabla \psi|_{\Sigma-}\}$$

We find β by solving (64) with (66) : it is an equation in one variable β. We can show that the lift C_f (the vertical component of the resultant of the force applied by the fluid on S) is proportional to β :

$$C_f = \beta \rho |u_\infty| \tag{67}$$

where ρ is the density of the fluid.

Practical Implementation :

In practice we can use P^1 finite elements though the Joukowski condition requires that we know $\nabla \varphi$ at the trailing edge ; with P^1, $\nabla \varphi$ is piecewise constant and so the triangles should be sufficiently small near the trailing edge. In [11] a rule for refining the triangles near the trailing edge (to keep the error $O(h)$:can be found the size of the triangles should decrease to zero as a geometric progression as they approach P and the rate of the geometric progression is a function of the angle of the trailing edge. Experience shows that one can apply the condition (61) by replacing P^+ and P^- by the triangles which are on S and have P as a vertex.

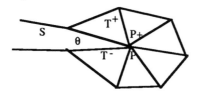

Figure 2.7 *Configuration of the triangles near the trailing edge.*

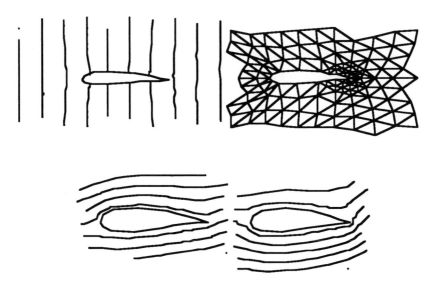

Figure 2.8 : *Example of numerical results : a) equipotentials
b) triangulation around the neighborhood of the profile
c) streamlines with lift d) and without lift.*

Another method, which is no doubt better from the theoretical point of view, but very costly in terms of computer programming time is to include a special basis function in place of the one associated with P, to represent the singularity. It can be shown that $\varphi(x)$ is like $|x-P|^{\pi/(2\pi-\beta)}$ in the neighborhood of the trailing edge, where β is the angle of the trailing edge. One can find the complete proof in Grisvard [102], and we give here only an intuitive justification.

Let us work in polar coordinates with origin at P and $\theta = 0$ corresponding to one edge of S and $\theta = 2\pi - \beta$ for the other. Let f be the limit of $\varphi(r,\theta)/r^m$ (m not necessarily an integer), i.e.

$$\varphi(r,\theta) = r^m f(\theta) + o(r^m)$$

The equations for φ are:

$$\frac{\partial \varphi}{\partial n}\Big|_{S^-} = r^{-1}\frac{\partial \varphi}{\partial \theta}(r,0) = 0 \Rightarrow \quad f'(0) = 0$$

$$\frac{\partial \varphi}{\partial n}|_{s+} = r^{-1}\frac{\partial \varphi}{\partial \theta}(r, 2\pi - \beta) = 0 \Rightarrow \quad f'(2\pi - \beta) = 0$$

$$\Delta \varphi = r^{-1}\frac{\partial}{\partial r} r \frac{\partial \varphi}{\partial r} + r^{-2}\frac{\partial^2 \varphi}{\partial \theta^2} = 0 \Rightarrow \quad m^2 f(\theta) + f''(\theta) = 0$$

The general solution of the last equation being $f(\theta) = a\sin(m(\theta - \theta^0))$ the first two conditions give :

$$\theta^0 = 0, \quad m(2\pi - \beta) = \pi$$

that is $m = \pi/(2\pi - \beta)$. We have therefore in the neighborhood of P:

$$\varphi(r, \theta) = a r^{\frac{\pi}{(2\pi - \beta)}} sin[\frac{\pi}{(2\pi - \beta)}\theta] + o(r^{\frac{\pi}{(2\pi - \beta)}})$$

which suggests taking a basis function associated with P as

$$w''(r, \theta) = r^{\frac{\pi}{(2\pi - \beta)}} sin[\frac{\pi}{(2\pi - \beta)}\theta]I_p(r)$$

where $I_p(r)$ is a smooth function which is equal to unity near P and zero far from P.

Results using this technique can be found in Dupuy [70], for example.

3.POTENTIAL SUBSONIC STATIONARY FLOWS :

3.1 Variational formulation

One still assumes that the effects of viscosity and thermal diffusion are negligible but does not assume that the fluid is incompressible. If the flow is stationary, irrotational at the boundary and behind all shocks if any, then one can solve the *transonic potential equation*

$$\nabla.[(k - \frac{1}{2}|\nabla\varphi|^2)^{\frac{1}{\gamma-1}}\nabla\varphi] = 0 \quad in \quad \Omega \tag{68}$$

The velocity is still given by

$$u = \nabla\varphi \tag{69}$$

the density by

$$\rho = \rho^0(k - \frac{1}{2}|\nabla\varphi|^2)^{\frac{1}{\gamma-1}}$$

and the pressure by

$$p = p^0(\frac{\rho}{\rho^0})^\gamma \tag{70}$$

The constants p^0, ρ^0 are usually known at the outer boundary. The boundary conditions are given on the normal flux $\rho u.n$ rather than on the normal velocity:

$$(k - \frac{1}{2}|\nabla\varphi|^2)^{\frac{1}{\gamma-1}}\frac{\partial\varphi}{\partial n} = g \quad on \quad \Gamma \tag{71}$$

To simplify the notation, let us put

$$\rho(\nabla\varphi) = \rho^0(k - \frac{1}{2}|\nabla\varphi|^2)^{\frac{1}{\gamma-1}} \tag{72}$$

Proposition 7
Problem (44)-(45) is equivalent to a variational equation. Find

$$\varphi \in W = \{w \in H^1(\Omega); \quad \int_\Omega \chi w = 0\}$$

(where χ is any given function with non zero mean on Ω) such that

$$\int_\Omega (k - \frac{1}{2}|\nabla\varphi|^2)^{\frac{1}{\gamma-1}}\nabla\varphi\nabla w = \int_\Gamma gw \quad \forall w \in W; \tag{73}$$

Moreover, any solution of (49) is a stationary point of the functional

$$E(\varphi) = -\int_\Omega (k - \frac{1}{2}|\nabla\varphi|^2)^{\frac{\gamma}{\gamma-1}} - \frac{\gamma}{\gamma-1}\int_\Gamma g \tag{74}$$

Proof :
To prove the equivalence between (68) and (73)-(74) we follow the procedure of proposition 1. Let us consider now

$$E(\varphi + \lambda w) =$$

$$-\int_\Omega (k - \frac{1}{2}|\nabla(\varphi + \lambda w)|^2)^{\frac{\gamma}{\gamma-1}} - \frac{\gamma}{\gamma-1}\int_\Gamma g(\varphi + \lambda w)$$

We have :

$$E'_{,\lambda}(\varphi + \lambda w) = \frac{\gamma}{\gamma-1}\int_\Omega (k - \frac{1}{2}|\nabla(\varphi + \lambda w|^2)^{\frac{1}{\gamma-1}}\nabla(\varphi + \lambda w)\nabla w - \frac{\gamma}{\gamma-1}\int_\Gamma gw$$

So all the solutions of (49) are such that

$$E'_{,\lambda}(\varphi + \lambda w)|_{\lambda=0} = 0$$

and now the result follows.

Proposition 8
If $b < (2k(\gamma-1)/(\gamma+1))^{1/2}$ then E defined by (70) is convex in

$$W^b = \{\varphi \in W : |\nabla\varphi| \le b\} \tag{75}$$

Proof :
Let us use (71) to calculate $E_{,\lambda\lambda}$ in the direction w :

$$\frac{\gamma - 1}{\gamma} \frac{d^2 E}{d\lambda^2}\Big|_{\lambda=0} = \int_\Omega \rho(\nabla\varphi)|\nabla w|^2 - \frac{1}{(\gamma - 1)} \int_\Omega \rho(\nabla\varphi)^{2-\gamma}(\nabla\varphi.\nabla w)^2$$

$$= \int_\Omega \rho(\nabla\varphi)^{2-\gamma}|\nabla w|^2(k - \frac{1}{2}|\nabla\varphi|^2[1 + \frac{2}{\gamma - 1}(\frac{\nabla\varphi.\nabla w}{|\nabla\varphi||\nabla w|})^2]$$

$$\geq \int_\Omega \rho(\nabla\varphi)^{2-\gamma}|\nabla w|^2(k - \frac{1}{2}|\nabla\varphi|^2(1 + \frac{2}{\gamma - 1}))$$

$$\geq (k - \frac{1}{2}b^2)^{\frac{2-\gamma}{\gamma-1}}(k - \frac{b^2}{2}\frac{\gamma + 1}{\gamma - 1})\|\nabla w\|_0^2$$

Corollary 1
If $b < (2k\,(\gamma - 1)/(\gamma + 1)\,)^{1/2}$ the problem

$$\min_{\varphi \in W^b} - \int_\Omega (k - \frac{1}{2}|\nabla\varphi|^2)^{\frac{\gamma}{\gamma-1}} - \frac{\gamma}{\gamma - 1}\int_\Gamma gw \qquad (76)$$

admits a unique solution.

Proof :
W^b is closed convex, E is $W - elliptic$, convex, continuous.

Corollary 2
If the solution of (76) is such that $|\nabla\varphi| \neq b$ at all points, then it is also the solution of (69).

3.2 . Discretisation

Proposition 9
Let $b < (2k\,(\gamma - 1)/(\gamma + 1)\,)^{1/2}$ and let

$$W_h^b = \{\varphi_h \in W_h : |\nabla\varphi_h| \leq b\} \qquad (77)$$

Assume that W_h^b and W^b satisfy (24) with strong convergence in $W^{1,\infty}$; let us approximate (75) by

$$\min_{\varphi_h \in W_h^b} - \int_\Omega (k - \frac{1}{2}|\nabla\varphi_h|^2)^{\frac{\gamma}{\gamma-1}} - \frac{\gamma}{\gamma - 1}\int_\Gamma g\varphi. \qquad (78)$$

If $\{\varphi_h\}_h$ are solutions in $W^{1,\infty}$ then $\varphi_h \to \varphi$ the solution of (76).

Proof :
As $|\nabla\varphi_h|$ is bounded by b, one can extract a subsequence which converges weakly in $W^{1,\infty}(\Omega)$ weak $*$. Let ψ be its limit.

Let φ be the solution of (76) and $\Pi_h\varphi$ the interpolate of φ in the sense of (24). Let E be the functional of problem (78). Then since $W_h^b \subset W^b$ we have :

$$E(\varphi) \leq E(\varphi_h) \leq E(\Pi_h\varphi).$$

The weak semi-continuity of E and the fact that φ_h is the solution of (78) implies that any element w of W^b is the limit of $\{w_h\}$, $w_h \in W_h^b$ in $W^{1,\infty}$ $(lim_{h\to 0}\|w - w_h\|_{1,\infty} = 0)$

$$E(\varphi) \leq E(\psi) \leq \liminf E(\varphi_h) \leq \lim E(\Pi_h\varphi) = E(\varphi) \qquad (79)$$

but φ is the minimum and so

$$E(\varphi) = E(\psi). \qquad (80)$$

The strong convergence of φ_h towards φ in $H^1(\Omega)$ is proved by using the convexity and the W^b -ellipticity of E.

3.3. Resolution by conjugate gradients :

To treat the constraint " $|\nabla\varphi| < b$ " the simplest method is penalization. We then solve :

$$\min_{\varphi_h \in W_h} E'(\varphi_h) \qquad (81)$$

$$E'(\varphi_h) = -\int_\Omega (k - \frac{1}{2}|\nabla\varphi_h|^2)^{\frac{\gamma}{\gamma-1}} - \frac{\gamma}{\gamma-1}\int_\Gamma g\varphi + \mu \int_\Omega [(b^2 - |\nabla\varphi_h|^2)^+]^2$$

where μ is the penalization parameter which should be large. The penalization is only to avoid the divergence of the algorithm if in an intermediate step the bound b is violated. If the solution reaches the bound b in a small zone then Corollary 2 doesn't apply any more; (experience shows that outside of this zone the calculated solution is still reasonable); but in this case it is better to use augmented Lagrangian methods (cf Glowinski [95]).

min: 101E-3   - x--x-: 120E-3   max: 273E-3

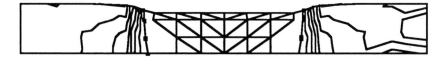

Figure 2.9: *Computed result of a transonic nozzle*
The figure represents the isovalues of $|\nabla\varphi|$, obtained by solving (81).

Taking into account the fact that W_h is of finite dimension, (81) is an optimization problem without constraint with respect to coefficients of φ_h on a basis of W_h :

$$\varphi_h(x) = \sum_1^{N-1} \varphi_i w^i(x)$$

$$\min_{\{\varphi_1 \cdots \varphi_{N-1}\}} E'(\varphi_h)$$

To solve (81), we use the conjugate gradient method with a preconditioning constructed from a Laplace operator with a Neumann condition. Let us recall the preconditioned conjugate gradient algorithm for the minimization of a functional.

Algorithm 2 (Preconditioned Conjugate Gradients) :

$$\text{Problem to be solved:} \quad \min_{z \in R^N} E(z) \tag{82}$$

0. Choose a preconditioning positive definite matrix C; choose $\epsilon > 0$ small, M a large integer. Choose an initial guess z^0. Put $n = 0$.

1. Calculate the gradient of E with respect to the scalar product defined by C; that is the solution of

$$Cg^n = -\nabla_z E(z^n) \tag{83}$$

If $\|g^n\|_C < \epsilon$ stop
     else if $n = 0$ put $h^0 = g^0$ else put

$$\gamma = \frac{\|g^n\|_C^2}{\|g^{n-1}\|_C^2} \tag{84}$$

$$h^n = g^n + \gamma h^{n-1} \tag{85}$$

3 . Calculate the minimizer ρ^n solution of

$$\min_\rho \{E(z^n + \rho h^n)\} \tag{86}$$

Put

$$z^{n+1} = z^n + \rho^n h^n \tag{87}$$

If $n < M$ increment n by 1 and go to 1 otherwise stop.

Proposition 10

If E is strictly convex and twice differentiable, algorithm 2 generates a sequence $\{z^n\}$ which converges ($\epsilon = 0$, $M = +\infty$) towards the solution of problem (82).

The proof can be found in Lascaux-Theodor [142] or Polak [193] for example. Let us recall that convergence is superlinear (the error is squared every N iterations) and that practical experiments show that $\sqrt{N}$ iterations are enough or even more than adequate when we have found a good preconditioner C. With a good preconditioner, the number of iterations is independent of N. This is the case with problem (81) when C is the matrix A constructed in (37). The linear systems (83) could be solved by algorithm 1. As in the linear case, the algorithm works even if the constraint on the mean of φ_h is active provided that the gradient g^n is projected in that space.

Gelder's Algorithm [86]:

The fixed point algorithm, introduced by Gelder [86], when it converges gives the result very fast :

$$\nabla.[(k - \frac{1}{2}|\nabla\varphi^n|^2)^{\frac{1}{\gamma-1}}\nabla\varphi^{n+1}] = 0 \quad in \quad \Omega \tag{88}$$

$$(k - \frac{1}{2}|\nabla\varphi^n|^2)^{\frac{1}{\gamma-1}}\frac{\partial\varphi^{n+1}}{\partial n} = g \quad on \quad \Gamma \tag{89}$$

It is conceptually very simple and easy to program. Once discretised by finite elements (88), (89) gives a linear system at each iteration, but the convergence is not guaranteed; however experience shows that it works well when the flow is everywhere subsonic. The coefficients depend on the iteration number n ; so one should avoid direct methods of resolution. If one solves (88) by algorithm 1 (conjugate gradient) then it becomes a method which from a practical point of view is very near to algorithm 2.

4. TRANSONIC POTENTIAL FLOWS :

4.1 Generalities :

If the solution of (75) is such that $|\nabla\varphi| = b$ almost everywhere on $I \subset \Omega$, measure(I) > 0 then we cannot solve (68) by (75).

From physics, we know that when the modulus of the velocity $|\nabla\varphi|$ is superior to $[2k(\gamma - 1)/(\gamma + 1)]^{1/2}$ the flow is supersonic. If we linearize (68) in the neighborhood of φ, we get :

$$\nabla.[\rho\nabla\delta\varphi - \frac{1}{\gamma - 1}\rho^{2-\gamma}\nabla\varphi\nabla\delta\varphi] = 0 \tag{90}$$

One can show by the same calculation as (74) that this equation is locally elliptic in the subsonic zone and locally hyperbolic in the supersonic zone. Furthermore numerical experience shows that (68) has many solutions in general C^0 but not C^1 ($|\nabla\varphi|$ is discontinuous), the discontinuities are where $|\nabla\varphi|$ is equal to the speed of sound. We have to impose an additional condition to avoid the jumps in the velocities (shocks) going from subsonic to supersonic in the flow direction.

We have seen in Chapter 1 that the entropy s is constant if η , ζ, κ are zero ; but this is no longer valid when the velocities are discontinuous ; when η is small, for example, $\eta\Delta u$ becomes big. It is shown in Landau-Lifchitz [141] that the entropy created by the shock is proportional to the third power of the jumps in the velocity across the shock in the flow direction : the entropy being increased through shocks it can only be produced by supersonic to subsonic shocks. An original method for imposing this condition (called an entropy condition) has been proposed by Glowinski [95] and studied by Festauer et al [77], Necas [180] :

Given 2 positive constants b and M, let

$$W_M^b = \{\varphi \in H^1(\Omega) : -\int_\Omega \nabla\varphi\nabla w \leq M \int_\Omega w \quad \forall w \geq 0 \tag{91}$$

$$w \in H^1(\Omega), \quad |\nabla\varphi| \leq b, \quad \int_\Omega \chi\varphi = 0\}$$

The variational inequality in (91) means that

$$\Delta\varphi \leq M \quad in \quad \Omega \tag{92}$$

which effectively avoids all positive jumps of $d\varphi/dx$ in the x-direction if ϕ is smooth in the y-direction.

Results leading towards the existence of solutions for (68)(71)(92) can be found in Kohn-Morawetz [135]. The following property is a key to show that subsequences converge in W_M^b.

Proposition 11
The space W_M^b is compact convex in $H^1(\Omega)$.

Proof (Festauer et al [77])
The convexity of W_M^b is clear ; let us show that it is compact. Let us define

$$G_n(w) = \int_\Omega (\nabla\varphi^n.\nabla w + Mw) \tag{93}$$

$$G(w) = \int_\Omega (\nabla\varphi\nabla w + Mw) \tag{94}$$

If $\varphi^n \to \varphi$ in $H^1(\Omega)$ weak* then $G_n \to G$ in $H^1(\Omega)$ weak* and as $G_n(w) \geq 0 \ \forall w$ we have, from a lemma of Murat [178], $G_n \to G$ strongly in $W^{1,\infty}(\Omega)$ weak*

But

$$|\nabla(\varphi^n - \varphi)|_o^2 = (G_n - G)(\varphi^n - \varphi) \to 0 \tag{95}$$

The functional used for the computation of subsonic flows is no longer convex in the supersonic zones; one must find another functional. A possibility is to minimize the square of the norm of the equation:

$$\min_{\varphi-\varphi_\Gamma \in H_0^1(\Omega)\cap W_M^b} ||\nabla.[(k - \frac{1}{2}|\nabla\varphi|^2)^{\frac{1}{\gamma-1}}\nabla\varphi]||_{-1}^2 \tag{96}$$

The choice of a suitable norm is very important to insure existence of solution and a fast algorithm. Here H^{-1} is clearly a good choice but it requires $\varphi|_\Gamma$ to be known (Dirichlet boundary conditions). With Neumann conditions (also with mixed conditions) one can solve:

$$\min_{\varphi \in W_M^b} \{ \int_\Omega |\nabla \epsilon|^2 dx : \tag{97}$$

$$(\nabla \epsilon, \nabla w) = (\rho(\nabla \varphi)\nabla \varphi, \nabla w) - \int_\Gamma gw \quad \forall w \in H^1(\Omega) \}$$

Indeed if we let ϵ be defined by

$$-\Delta \epsilon = \nabla \cdot [(k - \frac{1}{2}|\nabla \varphi|^2)^{\frac{1}{\gamma-1}} \nabla \varphi], \quad \epsilon \in H_0^1(\Omega)$$

then (96) can be written equivalently as

$$\min_{\varphi - \varphi_\Gamma \in H_0^1(\Omega) \cap W_M^b} \{ \int_\Omega |\nabla \epsilon|^2 \}.$$

A finite element discretisation of (97) is

$$\min_{\varphi_h \in W_{hM}^b} \{ \int_\Omega |\nabla \epsilon_h|^2 : \tag{98}$$

$$(\nabla \epsilon_h, \nabla w_h) = (\rho(\nabla \varphi_h)\nabla \varphi_h, \nabla w_h) - \int_\Gamma gw_h \quad \forall w_h \in W_h \}$$

It should be noted that this problem always has a solution since it is the minimization of a positive differentiable function in a finite dimensional bounded space. But φ_h will be an approximation to the solution of the transonic problem only if $\epsilon_h \approx 0$; and the set of points on which the constraints of W_{hM}^b are active must tend to a set of measure zero.

In Glowinski [95] details of the conjugate gradient solutions of this problem can be found. We refer to Lions [154] for a thorough study of optimal control problems.

4.2 . Numerical considerations

Practically, experience shows that the constraint (92) is always saturated and so the calculated solution depends on M (Cf. Bristeau et al [44]) and that the choice of M is thus critical.

In the actual computation one may prefer to use "upwinding" or "artificial viscosity" to take care of the entropy condition.

Upwinding will be studied in the following chapter. Let us give a simple version due to Hughes [115].

Equation (49) is approximated by

min: 580E-3　　- x--x-: 603E-3　　max: 789E-3

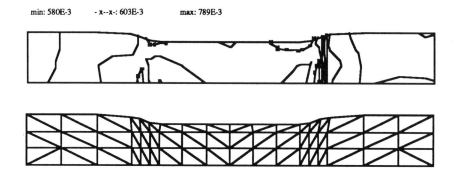

Figure 2.10: *Transonic flow computed by Gelder's algorithm with upwinding.*

$$\int_{\Omega}(k-\frac{1}{2}|\nabla\varphi_h^n|_d^2)^{\frac{1}{\gamma-1}}\nabla\varphi_h^{n+1}\nabla w_h = \int_{\Gamma}gw_h \quad \forall w_h \in W_h \quad \varphi^{n+1} \in W_h \quad (99)$$

where W_h is, for example, a P^1 conforming triangular finite element approximation of W and where $(\nabla\varphi_h)_d$ is calculated on the triangle $T_{j(k)}$, nearest to T_k "upwind" of the flow instead of calculating it on the triangle T_k

$$(\nabla\varphi_h^n)_d|_{T_k} = \nabla\varphi^n|_{T_{j(k)}}$$

Numerical results with this method can be found in Hughes [115] or Dupuy [70] for example.

Artificial viscosity will also be studied in chapter 3 ; one can add viscous terms to the equation (they were there in the Navier-Stokes equations and because of them the entropy can only grow). Following Jameson [121] (see also Bristeau et al.[44]), we consider

$$\int_{\Omega}(k-\frac{1}{2}|\nabla\varphi_h^n|^2)^{\frac{1}{\gamma-1}}\nabla\varphi_h^{n+1}\nabla w_h + h\int_{\Omega}[\frac{\partial}{\partial s}[(u'^{n2}-c^2)^+]\nabla\varphi_h^{n+1}\nabla\rho_h'^n]w_h\,dx$$

$$(100)$$

$$= \int_{\Gamma}gw_h$$

where the u'^n is an upwind approximation of $|\nabla\varphi_h^n|$, c is the speed of sound and $\rho_h'^n$ is an upwind approximation of $(k-(1/2)|\nabla\varphi_h^n|^2)^{1/\gamma-1}$; $\partial f/\partial s$ stands for $(u/|u|)\nabla f$, the derivative in the direction u ; this upwind approximation can be constructed as above or by

$$(\nabla\varphi_h^n)_d|_{T_k} = (1-\omega)\nabla\varphi_h^n|_{T_k} + \omega\nabla\varphi_h^n|_{T_{j(k)}}$$

where ω is a relaxation parameter.

Finally, notice that (99) and (100) are Gelder-type algorithms to solve underlying nonlinear equations; other methods can be used such as the H^{-1}

least-square method presented above ((96)-(98)) or GMRES, a quasi Newton algorithm which will be studied in chapter 5 ; results for these can be found in Bristeau et al [44].

5. VECTOR POTENTIALS :

Let us return to (1), (2) and study the possibility of calculating u by $u = \nabla \times \psi$. For simplicity, we shall assume that Ω is simply connected. Let Γ be piecewise twice differentiable and Γ_i be its simply connected components.

Proposition 12 :
Let $u \in L^2(\Omega)$ such that $\nabla.u \in L^2(\Omega)$. Let $\phi \in H^1(\Omega)/R$ the solution of

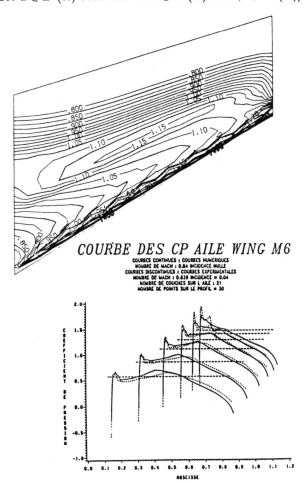

Figure 2.11: *Potential transonic flow around a M6 wing; computed by J.M. Dupuy: the figure shows the Mach numbers on the wings and presure coefficients on sections of the wing.*

$$(\nabla\phi, \nabla w) = (u, \nabla w) \quad \forall w \in H^1(\Omega)/R \tag{101}$$

and $\psi \in [H^1(\Omega)/R]^3$ the *solution of*

$$(\nabla \times \psi, \nabla \times v) + (\nabla.\psi, \nabla.v) = (u, \nabla \times v) \quad \forall v \in V \tag{102}$$

$$\psi \in V = \{v \in H^1(\Omega)^3 : v \times n|_\Gamma = 0, \int_{\Gamma_i} v.n = 0\}, \tag{103}$$

then

$$u = \nabla\phi + \nabla \times \psi \tag{104}$$

Proof :

The proof being rather long, we shall give only the main ideas and the details can be seen in Bernardi [30], Dominguez et al [67], Eldabaghi et al [73]. First let us note that (101) et (102) are well posed so that ϕ and ψ are unique. Then we see that the solution of (102) satisfies $\nabla.\psi = 0$. In fact, by taking v $= \nabla q$ in (102) with q zero on Γ (because of (103)),we see :

$$(\nabla.\psi, \Delta q) = 0 \quad \forall q \in H^2(\Omega) \cap H_0^1(\Omega) \ such \ that \ \int_\Gamma \frac{\partial q}{\partial n} = 0 \tag{105}$$

that is we almost have

$$(\nabla.\psi, g) = 0 \quad \forall g \in L^2(\Omega) \tag{106}$$

Nevertheless, one can prove, without much difficulty, that (105) implies $\nabla.\psi = 0$. Now let

$$\xi = \nabla \times \psi + \nabla\phi - u \tag{107}$$

we see that (101) implies

$$\nabla.\xi = 0 \quad \xi.n = n.\nabla \times \psi \tag{108}$$

and (102) implies

$$\nabla \times \xi = 0 \tag{109}$$

because of the following identity :

$$(a\nabla \times b) = (\nabla \times a, b) + \int_\Gamma a.(n \times b) \tag{110}$$

Besides, if τ_1, τ_2 are two vectors tangential to Γ and orthogonal to each other then

$$\psi \times n = 0 \quad \Rightarrow \psi.\tau_i = 0 \quad \Rightarrow (\psi.\tau_j)_{,\tau_j} = 0 \tag{111}$$

so that

$$n.\nabla \times \psi = -(\psi.\tau_1)_{,\tau_2} + (\psi.\tau_2)_{,\tau_1} = 0; \tag{112}$$

the proof is completed by using the following property :

$$\nabla \times \xi = 0, \quad \nabla.\xi = 0, \quad \xi.n|_\Gamma = 0 \Rightarrow \xi = 0. \tag{113}$$

Proposition 13 :
With the same conditions as in proposition 12, we can also use the following decomposition :
Let ϕ be the solution in $H^1(\Omega)/R$ of

$$(\nabla\phi, \nabla w) = -(\nabla.u, w) \quad \forall w \in H^1(\Omega)/R \tag{114}$$

and $\psi \in H^1(\Omega)^3$ the solution of

$$(\nabla \times \psi, \nabla \times v) + (\nabla.\psi, \nabla.v) = (u, \nabla \times v) \quad \forall v \in V \tag{115}$$

$$\psi \times n = \nabla q \text{ on } \Gamma, \int_{\Gamma_i} \psi.n = 0 \tag{116}$$

where q is the solution of the Beltrami equation on Γ.

$$\sum_{i=1,2} \int_\Gamma \frac{\partial q}{\partial \tau_i} \frac{\partial w}{\partial \tau_i} = \int_\Gamma u.nw \quad \forall w \in H^1(\Gamma) \tag{117}$$

where τ_i are two orthogonal tangents to Γ. Then

$$u = \nabla \times \psi + \nabla\varphi$$

Proof :
The proof is the same as that of proposition 12, except for (111) onward. Instead of (112), we have:

$$n.\nabla \times \psi = -(\frac{\partial^2 q}{\partial \tau_1^2} + \frac{\partial^2 q}{\partial \tau_2^2}) = u.n \tag{118}$$

where the last equality is due to (117). Finally, we also have $\xi.n = 0$ and so $\xi = 0$.

We have therefore two decompositions of u. In the first one, $\partial\phi/\partial n = u.n$ whereas in the second $n.\nabla \times \psi = u.n$ and $\partial\phi/\partial n = 0$.

If $\nabla.u = 0$ in order that $\phi \equiv 0$, one has to use the second decomposition. So we see that the potential vector resolution of incompressible 3D flow requires, in general,

- the solution of a Beltrami equation on the boundary (117),
- the solution of a vectorial elliptic equation (115).

Remark

An alternative form for (117) is

$$\int_{\Gamma} \nabla q \nabla w - \frac{\partial q}{\partial n} \frac{\partial w}{\partial n} = \int_{\Gamma} u.nw, \forall w \in H^1(\Gamma)$$

6. ROTATIONAL CORRECTIONS:

We can use proposition 12 and the transonic equation (64) to construct a method of solving the stationary Euler equations :

$$\nabla.(\rho u) = 0 \tag{119}$$

$$\nabla.(\rho u \otimes u) + \nabla p = 0 \tag{120}$$

$$\nabla.[\rho u(\frac{1}{2}u^2 + \frac{\gamma}{\gamma - 1}\frac{p}{\rho})] = 0 \tag{121}$$

As in chapter 1, we deduce from (121) that $p\rho^{-1}\gamma/(\gamma - 1) + u^2/2$ is constant on the streamlines, that is, if x_∞ is the upstream intersection of the stream line with Γ, we have

$$\frac{\gamma}{\gamma - 1}\frac{p}{\rho} + \frac{u^2}{2} = H(x_\infty) \text{ on } \{x : x' = u(x), \quad x(0) = x_\infty\} \tag{122}$$

By definition of the reduced entropy S, we have

$$\frac{p}{\rho^\gamma} = S \tag{123}$$

and S is constant on the streamlines, except across the shocks. In fact, $\omega = \nabla \times u$ satisfies

$$0 = \nabla.(\rho u \otimes u) + \nabla p = -\rho u \times \nabla \times u + \nabla p + \rho \nabla \frac{u^2}{2} \tag{124}$$

$$= -\rho u \times \omega + \rho \nabla H - \frac{\rho^\gamma}{\gamma - 1}\nabla S.$$

and so $\rho u \nabla S = 0$, but this calculation is not valid if u is discontinuous. However we can deduce from (124) that

$$\omega = (u \times \nabla H - \frac{\rho^{\gamma-1}}{\gamma - 1}u \times \nabla S)|u|^{-2} + \lambda \rho u \tag{125}$$

where λ is adjusted in such a way that $\nabla.\omega = 0$, i.e.

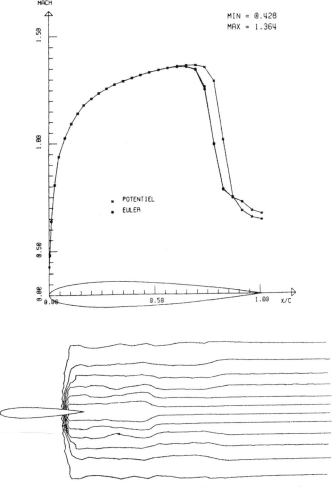

Figure 2.12: *Transonic flow with entropy correction; computed by F. El-Dabaghi using the Helmoltz decomposition. The Mach numbers are shown with and without entropy corection; the other figure shows the entropy.*

$$\rho u \nabla \lambda = -\nabla.[(u^{-2})(u \times \nabla H - \rho^{\gamma-1} u \times \frac{\nabla S}{\gamma-1})] \tag{126}$$

Let us now use proposition 12 ; we get the following algorithm :

Algorithm

O. Choose u continuous and calculate the streamlines

1. Calculate a new H by transporting $H(x_\infty)$ on the streamlines and a new S by transporting $S(x_\infty)$ on the streamlines and if necessary adding a jump $[S]$ calculated from the Rankine Hugoniot conditions across the shocks (conservation of $[\rho u]$, $[\rho u^2 + p]$, $[u^2/2 + \gamma p/(\gamma-1)\rho]$).

2. Calculate ω by (125)with ρ calculated by (122) and (123), i.e.

$$\rho(u) = \frac{\gamma - 1}{\gamma S}(H - \frac{1}{2}u^2)]^{\frac{1}{\gamma - 1}} \tag{127}$$

and λ calculated by (126).

3 . Calculate ψ by resolving (102) (we shall apply (110) to calculate the right hand side).

4 . Calculate ϕ by solving (119), i.e. with (127) :

$$\nabla.[\rho(\nabla\phi + \nabla \times \psi)] = 0 \quad \rho\frac{\partial\phi}{\partial n}|_\Gamma = g \tag{128}$$

5 . Put $u = \nabla\phi + \nabla \times \psi$ and return to 1.

We remark that when $\omega << 1$, ψ is small so all that remains is (128) i.e. the transonic equation. This algorithm is therefore a rotational correction to the transonic equation. The convergence of the algorithm is an open problem but it seems to be quite stable (see El Dabaghi et al [73]) and it gives good results when ω is not too large. From a practical point of view, the main difficulty remains in the identification of shocks and applying the Rankine-Hugoniot condition in step 1 (see Hafez et al [105] or Luo [161] for an alternative method).

Finally, we see that it requires the integration of 3 equations of the type :

$$u\nabla\xi = f, \quad \xi|_\Sigma \text{ given}; \tag{129}$$

this will be treated in the next chapter.

Other decompositions have been used to solve the Euler equations. An example is the Clebsch decomposition used by Aker et al. [71] and Zijl [242] $u = \nabla\varphi + s\nabla\psi$ where φ, s, ψ are scalar valued functions. Extension of these methods to the Navier-Stokes equations can be found in El Dabaghi et al [73], Ecer et. al. [71], Zijl [242].

Chapter 3

Convection diffusion phenomena

1. INTRODUCTION

The following PDE for ϕ is called a linear convection-diffusion equation:

$$\phi_{,t} + a\phi + \nabla.(u\phi) - \nabla.(\nu\nabla\phi) = f \tag{1}$$

where $a(x,t)$ is the dissipation coefficient (positive), $u(x,t)$ is the convection velocity and $\nu(x,t)$ is the diffusion coefficient (positive).

For simplicity we assume that $a = 0$, as is often the case in practice. All that follows, however, applies to the case $a > 0$; for one thing when a is constant we have:

$$(e^{at}\phi)_{,t} = e^{at}(\phi_{,t} + a\phi)$$

so with such a change of function one can come back to the case $a = 0$; secondly the diffusion terms and the dissipative terms have similar effects physically.

These equations arise often in fluid mechanics. Here are some examples :
- Temperature equation for θ for incompressible flows,
- Equations for the concentration of pollutants in fluids,
- Equation of conservation of matter for ρ and the momentum equation in the Navier-Stokes equations with $\phi = u$ although these equations are coupled with other equations in which one could observe other phenomena than that of convection-diffusion.

In general, ν is small compared to UL (characteristic velocity × characteristic length) so we must face two difficulties:
  - boundary layers,
  - instability of centered schemes.

On equation (1) in the stationary regime when $\nu = 0$ we shall study first three types of methods:
  - characteristic methods,
  - streamline upwinding methods,
  - upwinding-by-discontinuity.
Then we will study the full equation (1) beginning with the centered schemes and finally for the full equation, the three above mentioned methods plus the Taylor-Galerkin / Lax-Wendroff scheme.

1.1. Boundary layers

To demonstrate the boundary layer problem, let us consider a one dimensional stationary version of (1) :

$$u\phi_{,x} - \nu\phi_{,xx} = 1 \tag{2}$$

$$\phi(0) = \phi(L) = 0 \tag{3}$$

where u and ν are constants. The analytic solution is

$$\phi(x) = (\frac{L}{u})[\frac{x}{L} - \frac{e^{[(uL/\nu)(x/L)]} - 1}{e^{(uL/\nu)} - 1}] \tag{4}$$

When $\nu/uL \to 0$,

$$\phi \to \frac{(x - L)}{u} \text{ if } u < 0, \text{ except in } x = 0,$$

$$\phi \to \frac{x}{u} \text{ if } u > 0, \text{ except in } x = L.$$

But the solution of (2) with $\nu = 0$ is

$$\phi = \frac{(x - L)}{u} \tag{5}$$

if we take the second boundary condition in (3) and

$$\phi = \frac{x}{u} \tag{6}$$

if we take the first one.

We see that when $\nu/uL \to 0$, the solution of (2) tends to the solution with $\nu = 0$ and one boundary condition is lost; at that boundary, for $\nu << 1$, there is a boundary layer and the solution is very steep and so it is difficult to calculate with a few discretisation points.

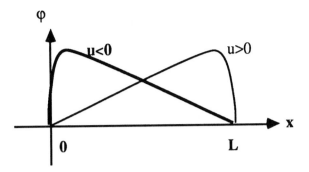

Figure 3.1 : *According to the sign of u, the solution is given by (5) or (6) with a boundary layer to catch the violated boundary condition.*

1.2. Instability of centered schemes :

Finite element methods correspond to centered finite difference schemes when the mesh is uniform and these schemes do not distinguish the direction of flow.

Let us approximate equation (2) by P^1 conforming finite elements. The variational formulation of (2)

$$\int_0^L u\phi_{,x}w + \nu \int_0^L \phi_{,x}w_{,x} = \int_0^L w, \quad \forall w \in H_0^1(]0,L[) \qquad (7)$$

is approximated by

$$\int_0^L (u\phi_{h,x}w_h + \nu\phi_{h,x}w_{h,x} - w_h) = 0 \quad \forall w_h \text{ continuous } P^1 \text{ piecewise} \qquad (8)$$

and zero on the boundary.

If $]0,L[$ is divided into intervals of length h, it is easy to verify that (8) can be written as

$$\frac{u}{2h}(\phi_{j+1} - \phi_{j-1}) - \frac{\nu}{h^2}(\phi_{j+1} - 2\phi_j + \phi_{j-1}) = 1, \quad j = 1, ..., N-1 \qquad (9)$$

where $\phi_j = \phi_h(jh)$, $j = 0,...,N = L/h$; $\phi_o = \phi_N = 0$.

Let us find the eigenvalues of the linear system associated with (9), that is, let us solve

$$I_n(\lambda) = det \begin{pmatrix} 2\alpha - \lambda & 1 - \alpha & 0 & ... \\ -1 - \alpha & 2\alpha - \lambda & 1 - \alpha & ... \\ ... & ... & ... & ... \\ ... & -1 - \alpha & 2\alpha - \lambda & 1 - \alpha \\ ... & & -1 - \alpha & 2\alpha - \lambda \end{pmatrix} = 0$$

where we have put $\alpha = 2\nu/(hu)$. One can easily find the recurrence relation for I_n :

$$I_n = (2\alpha - \lambda)I_{n-1} + (1 - \alpha^2)I_{n-2}$$

and so when $\lambda = 2\alpha$ then $I_{2p+1} = 0 \; \forall p$. This eigenvalue is proportional to ν, and tends to 0 when $\nu \to 0$. So one could foresee that the *system is unstable when ν/hu is very small.*

There are several solutions to there difficulties :

1. One can try to remove the spurious modes corresponding to null or very small eigenvalues by putting constraints on the finite element space (see Stenberg [220] for example).

2. One can solve the linear systems by methods which work even for non-definite systems. For example Wornom- Hafez [239] have shown that a block relaxation method with a sweep in the direction of the flow allows (9) to be solved without upwinding.

3. Finally one can modify the equation or the numerical scheme so as to obtain non-singular well posed linear systems: This is the purpose of upwinding and artificial diffusion.

Many upwinding schemes have been proposed (Lesaint-Raviart [149], Henrich et al. [110], Fortin-Thomasset [83], Baba-Tabata [7], Hughes [115], Benque et al. [20], Pironneau [190] ... see Thomasset [229] for example). We shall study some of them. For clarity we begin with the stationary version of (1).

2. STATIONARY CONVECTION:

2.1. Generalities :

In this section, we shall study the problem

$$\nabla.(u\phi) = f \text{ in } \Omega \quad \phi|_\Sigma = \phi_\Gamma \tag{10}$$

where u is given in $W^{1,\infty}(\Omega)$, f is given in $L^1(\Omega)$ and

$$\Sigma = \{x \in \Gamma : n(x).u(x) < 0\} \tag{11}$$

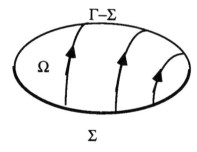

Figure 3.2 : Σ *is the part of* Γ *where the fluid enters* ($u.n < 0$).

A variational formulation for this problem is to search for $\phi \in L^2(\Omega)$ such that

$$(\phi, u\nabla w) = \int_\Sigma w\phi_\Gamma u.n - \int_\Omega fw \quad \forall w \in H^1(\Omega) \quad with \quad w_{\Gamma-\Sigma} = 0.$$

Here $\phi_\Gamma \in H^{1/2}(\Sigma), u, \nabla.u \in L^2(\Omega), f \in L^2(\Omega)$ is enough but with more regularity there is an explicit solution:

Proposition 1 :
Let $X(x;s)$ be the solution of

$$X' = u(X), \quad X(0) = x. \tag{12}$$

Let $X_\Gamma(x) = X(x; s_\Gamma)$ be the intersection of $\{X(x;s) : s < 0\}$ with Σ.
Then

$$\phi(x) = \phi_\Gamma(X_\Gamma(x)) e^{-\int_{s_\Gamma}^0 \nabla.u(X(x;s))ds} + \int_{s_\Gamma}^0 f(X(x;s)) e^{\int_s^0 -\nabla.u(X(x;\tau))d\tau} ds \tag{13}$$

is solution of (10).

Proof :

$$u.\nabla\phi = X'\nabla\phi = \frac{d}{ds}\phi(X(x;s)) = f(X(x;s)) - \phi\nabla.u$$

on integration this equation gives (13).

Corollary 1 :
If all the streamlines intersect Σ, then (10) has a unique solution.

Remark :
The solution of (10) evidently has the following properties :
- positivity : $f \geq 0, \phi_\Gamma \geq 0 \Rightarrow \phi \geq 0$
- conservativity : $\nabla.u = 0, f = 0 \Rightarrow \int_S u.n\phi ds = 0 \,\forall S$ closed contour $\subset \Omega$
- stability : $|u\phi|(x) \leq |f|_{L^1(\Omega)} - u.n(X_\Gamma(x))|\phi_\Gamma(X_\Gamma(x))|$
We seek, if possible, numerical schemes which preserve the above properties.

A priori the simple P^1 triangular conforming finite elements and

$$(w_h, \nabla.(u\phi_h)) = (f, w_h) \quad \forall w_h \in W_h \quad w_h|_\Sigma = 0; \quad \phi_h \in W_h \quad \phi_h|_\Sigma = 0 \tag{14}$$

do not preserve the above properties. In fact, if Ω is a square divided into triangles with the sides parallel to the axes and if $u = (1,0)$ we obtain the centered finite difference scheme, which we know is unstable, when ϕ is irregular:

$$2(\phi_{i+1,j} - \phi_{i-1,j}) + \phi_{i+1,j+1} - \phi_{i,j+1} + \phi_{i,j-1} - \phi_{i-1,j-1} = 6hf_{i,j}$$

where $\phi_{i,j}$ is the value of ϕ at vertex ij. The solution of (14) cannot be unique when $\nabla.u = 0$ since it is a skew symmetric linear system .

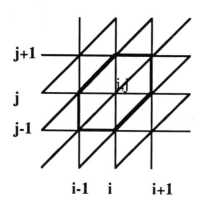

Figure 3.3: *A uniform triangulation and the corresponding numbering.*

2.2. Scheme 1 (characteristics) :

One could propose the following scheme :
1) u is approximated by u_h , P^1 conforming, from the values of u at the vertices q^i.
2) The characteristics $\{X(q^i;.)\}$ originating from the nodes $\{q^i\}$ are approximated by polygonal lines $\cup_k [\xi^k \ \xi^{k+1}]$, $\xi^0 = q^i$, where ξ^{k+1} is the intersection of the line $\{\xi^k - \mu u_h(\xi^k)\}_{\mu>0}$ with the boundary ∂T_l of the triangle T_l which contains ξ^k and $\xi^k - \epsilon u_h(\xi^k)$, ϵ positive, small (see figure 3.4).
3) (13) is evaluated at x = q^i and this defines $\phi(q^i)$
4) $\phi_h(x)$ is defined from $\phi(q^i)$ by a P^1 interpolation.

Proposition 2 :

$$|\phi - \phi_h|_\infty \le hC \tag{15}$$

where C depends on $|\phi_\Gamma|_{1,\infty}, |f|_{1,\infty}, |u|_{1,\infty}, |f|_\infty$ and on $diam(\Omega)$.

Proof :
To simplify, let us assume that $\nabla.u = 0$; the reader can build a similar proof in the general case.
Let us evaluate the error along the characteristics; let $X_h(x;.)$ be the solution of

$$X' = u_h(X), X(0) = x,$$

approximated as in step 2 above ; then

$$|X'_h - X'| = |u_h(X_h) - u(X)| \leq |(u_h - u)(X_h)| + |u(X_h) - u(X)|$$

$$\leq |u_h - u|_\infty + |\nabla u|_\infty |X_h - X|$$

Using the Belman-Gronwall lemma :

$$|X_h - X| \leq |u_h - u|_\infty (e^{|\nabla u|_\infty L} - 1) \tag{16}$$

where L is the length of the longest characteristic. Finally,

$$|\phi_h(q^i) - \phi(q^i)| \leq |\phi_\Gamma(X_{\Gamma h}(q^i)) - \phi_\Gamma(X_\Gamma(q^i))| + |\int_0^{s_{\Gamma h}} f(X_h)ds - \int_0^{s_\Gamma} f(X)ds|$$

$$\leq |\nabla \phi_\Gamma|_\infty |(X_{\Gamma h} - X_\Gamma)(q^i)| + L|\nabla f|_\infty |X_h - X|_\infty + |s_{\Gamma h} - s_\Gamma||f|_\infty \tag{17}$$

As $|s_{\Gamma h} - s_\Gamma|$ is the difference in length between the discrete and the continuous characteristics , so long as the boundary is regular enough with the help of (16) we obtain the result:

$$|\phi_h(q^i) - \phi(q^i)|_\infty \leq C|u - u_h|_\infty$$

Numerical considerations :
    This scheme is accurate but rather costly because one has to calculate all the characteristics starting from all the vertices. For example for a square domain with N^2 vertices, one must cross $N^2(N-1)$ triangles in total when the velocity is horizontal, so the cost of this computation is $0(N_s^{3/2})$ if N_s denotes the total number of vertices.
    One could reduce the cost by first calculating them for the q^i which are far from Σ in the direction u then using a linear interpolation for all the nodes for which there exists two opposite triangles already crossed by previously computed characteristics.

Remark (Saiac)
    If $u = \nabla \times \psi$ and ψ is known then $X_{\Gamma h}$ can be computed from the equation $\psi(X_{\Gamma h}(q^i)) = \psi(q^i)$, $X_{\Gamma h}(q^i) \in \Gamma$; so there is no need to cross the domain starting from q^i.

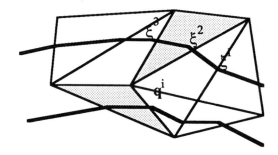

Figure 3.4: *The characteristics are broken lines joining vertices ξ^i. If two characteristics go around a vertex q, then the characteristic which starts from q may be skipped.*

To calculate the $\{\xi^k\}_k$ a good algorithm is to use the barycentric coordinates $\{\lambda_j^{k+1}\}_{j=1..m+1}$ of ξ^k and decompose $u_h(\xi^k)$ in μ_j^k such that

$$\sum_j \mu_j^k q^j = u_h(\xi^k) \tag{18}$$

$$\sum_j \mu_j^k = 0 \tag{19}$$

then $\xi^{k+1} = \sum_{j=1..n} \lambda_j^{k+1} q^j$ $(n = 2$ or $3)$ with

$$\lambda_j^{k+1} = \rho \mu_j^k + \lambda_j^k \text{ where } \rho \text{ is such that}$$

$$\lambda_j^{k+1} \geq 0, \quad \exists l \text{ such that } \lambda_l^{k+1} = 0.$$

This calculation is explained in detail in (120).

2.3. Scheme 2 (upwinding by discretisation of the total derivative)

A weak formulation of (10) is written as:
Find ϕ in $L^2(\Omega)$ such that

$$-(\phi, u\nabla w) + \int_\Sigma \phi_\Gamma wu.n = (f, w) \quad \forall w \in H^1(\Omega), \text{ such that } w|_{\Gamma-\Sigma} = 0.$$

Let $T_\delta^\pm(\Omega)$ be a translation of Ω by δ in the direction $\pm u$

$$T_\delta^\pm(\Omega) = \{x \pm u(x)\delta : \quad x \in \Omega\}$$

To construct the scheme, we consider the following two approximations :

$$u\nabla w(x) \cong \frac{1}{\delta}[w(x + u(x)\delta) - w(x)]$$

$$\int_{\Omega - \Omega \cap T_\delta^-(\Omega)} g = \delta \int_{\Gamma - \Sigma} gu.n + o(\delta)$$

With these approximations (20) can be approximated in the space W_h of P^1 functions continuous on the triangulation Ω by the following problem :

$$(\phi_h, w_h) - \int_{\Omega \cap T_\delta^-(\Omega)} \phi_h(x) w_h(x + u(x)\delta) = \delta(f, w_h) - \delta \int_\Sigma \phi_\Gamma w_h u.n,$$

$$\forall w_h \in W_h; w_h|_{\Gamma - \Sigma} = 0$$

or, with the convention that $w_h(x) = 0$ if $x \notin \Omega$:

Find $\phi_h \in W_h$ such that for all $w_h \in W_h$ with $w_h|_{\Gamma - \Sigma} = 0$:

$$(\phi_h, w_h) - \int_\Omega \phi_h(x) w_h(x + u(x)\delta) dx = \delta(f, w_h) - \delta \int_\Sigma \phi_\Gamma w_h u.n,$$

One may prefer schemes where the boundary conditions are satisfied in the strong sense. For this one should start with the following variational formulation of (10):

Find ϕ such that $\phi|_\Sigma = \phi_\Gamma$ and

$$-(\phi, u\nabla w) + \int_{\Gamma - \Sigma} \phi w u.n = (f, w) \quad \forall w \in H^1(\Omega), \quad w|_\Sigma = 0. \qquad (20)$$

Then by the same argument one finds the following discrete problem:

Find ϕ_h with $\phi_h - \phi_{\Gamma_h} \in W_h$ and

$$(\phi_h, w_h) - \int_{\Omega \cap T_\delta^-(\Omega)} \phi_h(x) w_h(x + u(x)\delta) dx = \delta(f, w_h), \quad \forall w_h \in W_h \qquad (21)$$

with $W_h = \{w_h \in C^0(\Omega) : w_h|_{T_k} \in P^1 \ \forall k, \ w_h|_\Sigma = 0 \}$.

Comments :
1. We could have performed the same construction by using the formula :

$$\nabla.(u\phi) = u\nabla\phi + \phi\nabla.u \cong \frac{1}{\delta}(\phi(x) - \phi(x - u(x)\delta)) + \phi(x - u(x)\delta)\nabla.u(x)$$

and could have obtained, instead of (21),

$$(\phi_h, w_h) = \int_\Omega \phi_h(x - u(x)\delta) w_h(x)(1 - \delta\nabla.u(x)) dx + \delta(f, w_h), \quad \forall w_h \in W_h. \qquad (22)$$

This is also a good scheme, but as will become clearer the scheme (21) is conservative whereas (22) is not.

2. If we use mass lumping (a special quadrature formula with the Gauss points at the vertices), that is, $\{q^i\}_i$ being the vertices of T:

$$\int_T f \cong \frac{|T|}{(n+1)} \sum_{1..n+1} f(q^i) \tag{23}$$

where $n = 2$ (resp.3), $|T|$ the area (resp. volume) of a triangle (resp. tetrahedron) (this formula is exact if f is affine) then the linear system from (21) and (22) becomes quasi-tridiagonal, in particular (22) becomes :

$$\phi_h(q^i) = \phi_h(q^i - u(q^i)\delta)(1 - \delta\nabla.u(q^i)) + f(q^i)\delta \tag{24}$$

On a uniform grid when u is constant, this is Lax's scheme [144] ; it is a positive but of order 1; however, the finite element scheme (21) is somewhat better (the dissipation comes mostly from the mass lumping in the second integral) but it is more complicated because it requires the solution of a non-symmetric linear system and calculation of the integral of a product of two P^1 functions on different grids (an analysis of these methods can be found in Bermudez et al [28]).

We also see how one can construct schemes of higher order by using a better approximation of $u\nabla w$.

Proposition 3 :
If u is constant, and if all the streamlines cut Σ, problem (21) has a unique solution

Proof :
If there were 2 solutions, their difference ϵ_h would be zero on Σ and would satisfy:

$$\int_\Omega \epsilon_h w_h - \int_{\Omega\cap T_\delta^-(\Omega)} \epsilon_h w_h o(I + u\delta) = 0 \tag{25}$$

Taking $w_h = \epsilon_h$, we would obtain :

$$|\epsilon_h|^2_{0,\Omega} \leq |\epsilon_h|_{0,\Omega} |\epsilon_h o(I + u\delta)|_{0,\Omega\cap T_\delta^-(\Omega)} \tag{26}$$

but

$$|\epsilon_h o(I + u\delta)|^2_{0,\Omega\cap T_\delta^-(\Omega)} = \int_{\Omega\cap T_\delta^+(\Omega)} \epsilon_h(y)^2 det|\nabla(x + u\delta)^{-1}|dy \tag{27}$$

$$= |\epsilon_h|^2_{0,\Omega\cap T_\delta^+(\Omega)}$$

So

$$|\epsilon_h|_{0,\Omega-\Omega\cap T_\delta^+(\Omega)} = 0 \qquad (28)$$

which means that ϵ_h is zero on all the vertices of the triangles which have a non-empty intersection with $\Omega - \Omega \cap T_\delta^+(\Omega)$.

Applying similar reasoning to $\Omega \cap T_\delta^+(\Omega)$ and proceeding step by step, we can show that ϵ_h is zero everywhere.

Remark on the convergence :

At least for the case u constant, using the same technique ,we would get an error estimate of the type

$$|\phi_h - \phi|_{0,\Omega-\Omega\cap T_\delta^+(\Omega)} \leq C(\delta^2 + h^2)$$

and so the final error is $0(\delta+h^2/\delta)$. The diffusivity of the schemes is a problem; Bristeau-Dervieux [40] have studied a 2nd order interpolation of $u\nabla w$ which gives better results, i.e. less dissipations, (but which is less stable).

Generalization of the previous schemes :

The first scheme corresponds to an integration along the whole length of the characteristics whereas the previous scheme corresponds to a repeated integration on a length δ. If one wants an intermediate method, one could proceed as follows (Benque et al. [20]) :

Let D be a subset of Ω and $\psi_w(x)$ the solution of

$$u\nabla\psi_w = 0 \text{ in } D, \quad \psi_w|_{\partial - D} = w$$

where

$$\partial^- D = \{x \in \partial D : u(x).n(x) < 0\}$$

As for (13), it is easy to see that

$$\psi_w(x) = w(X_{\partial D}(x))$$

where X is the solution of (12) and $X_{\partial D}$ the intersection with $\partial^- D$. From (10) we deduce

$$\int_D f\psi_w = \int_D \nabla.(u\phi)\psi_w = \int_{\partial - D} w\phi u.n + \int_{\partial + D} \psi_w\phi u.n$$

where ∂D^+ is the part of ∂D where u.n ≥ 0. By choosing an approximation space W_h, a method to calculate $X_{\partial D}$ and a quadrature formula, one could define a family of schemes of the type :

$$\int_{\partial + D} \phi_h u.n w_h o X_{h\partial D} + \int_{\partial - D} w_h \phi_h u.n = \int_D f(x)w(X_{h\partial D}(x))dx \quad \forall w_h \in W_h$$

and this family contains the above two schemes.

2.4. Scheme 3 (Streamline upwinding SUPG) :

The streamline upwinding scheme (also called SUPG=Streamline Upwinding Petrov Galerkin) was proposed by Hughes [116] and studied by Johnson et al. [125]. On the space W_h of P^1 functions continuous on the triangulation of Ω and zero on Σ, for example, one searches ϕ_h such that $\phi_h - \phi_{\Gamma h} \in W_h$ with

$$(\nabla.(u\phi_h) - f, w_h + h\nabla.(uw_h)) = 0 \quad \forall w_h \in W_h \qquad (29)$$

Comments :
The idea is the following :
The finite element method leads to centered finite difference schemes because the usual basis functions w^i (the "hat function") are symmetric (on a uniform triangulation) with respect to vertices q^i. So replacing the Galerkin formulation

$$(\nabla.(u\phi_h), w^i) = (f, w^i) \quad \forall i, \quad \phi_h = \phi_{\Gamma h} + \sum \phi_i w^i \qquad (30)$$

by a nonsymmetric "Petrov-Galerkin" formulation (cf. Christie et al. [54])

$$(\nabla.(u\phi_h), w'^i) = (f, w'^i) \qquad (31)$$

where the w'^i have more weight upstream than downstream ; that is the case of the function $x \to w^i + h\nabla.(uw^i)$ (see figure 3.5). We note also that one could work either with w'^i or w^i in the second member.

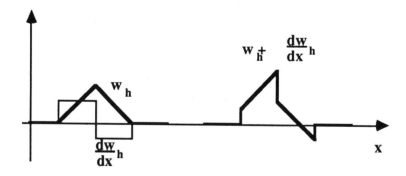

Figure 3.5 :
The P^1 basis functions and the Petrov-Galerkin functions of Hughes.

Another interpretation could also be given ;
Equation (29) is also an approximation of

$$\nabla.(u\phi) - hu\nabla(\nabla.(u\phi)) = f - hu\nabla f \qquad (32)$$

that is, when u is constant

$$\nabla.(u\phi) - h\nabla.[(u \otimes u)\nabla\phi] = f - hu\nabla f \qquad (33)$$

So we have in fact added a tensorial viscosity $hu \otimes u$; now this tensor has its principal axis in the direction u, that is we have added viscosity only in the direction of the flow. Furthermore what is added on the left is also added on the right because (10) differentiated gives $u\nabla(\nabla.[u\phi]) = u\nabla f$.

Proposition 4 :
i) *The scheme (29) has a unique solution*
ii) *If $u \in W^{1,\infty}$ the solution satisfies the following inequality :*

$$|\phi_h - \phi|_{0,\Omega} \le ch^{\frac{3}{2}}|\phi|_{2,\Omega} \qquad (34)$$

Proof when $\nabla.u \ge \alpha > 0$:
If (29) has two solutions, their difference ϵ_h satisfies

$$h|\nabla.(u\epsilon_h)|_0^2 + \frac{1}{2}\int_\Omega (\nabla.u)\epsilon_h^2 + \frac{1}{2}\int_{\Gamma-\Sigma} u.n\epsilon_h^2 d\gamma = 0 \qquad (35)$$

because

$$(\nabla.(u\epsilon_h), \epsilon_h) = \frac{1}{2}((\nabla.u)\epsilon_h, \epsilon_h) + \frac{1}{2}\int_\Gamma u.n\epsilon_h^2 \qquad (36)$$

So ϵ_h is zero.

Let us find an error estimate :

Subtracting the exact equation from the approximated equation, we obtain:

$$(\nabla.(u(\phi - \phi_h)), w_h + h\nabla.(uw_h)) = 0 \quad \forall w_h \in W_h \qquad (37)$$

By taking $w_h = \psi_h - \phi_h$ where ψ_h is the interpolation function of ϕ, we get :

$$(\nabla.(uw_h), w_h + h\nabla.(uw_h)) = (\nabla.(u(\psi_h - \phi)), w_h + h\nabla.(uw_h)) \qquad (38)$$

$$= ((\nabla.u)(\psi_h - \phi), w_h) - (\nabla.(uw_h), \psi_h - \phi - h\nabla.(u(\psi_h - \phi)))$$

$$+ \int_{\Gamma-\Sigma} u.nw_h(\psi_h - \phi)$$

or using (36) and the error estimate of interpolation ($|\phi - \psi_h|_\alpha \le ch^{2-\alpha}$, $\alpha = 0, 1$ and $|\phi - \psi_h|_{0,\Gamma} \le ch^{3/2}$):

$$\frac{1}{2}\int_\Omega (\nabla.u)w_h^2 + \frac{1}{2}\int_{\Gamma-\Sigma} u.nw_h^2 + h|\nabla.(uw_h)|_0^2$$

$$\le Ch^2|\nabla.(uw_h)|_0 + h^{\frac{3}{2}}|u.nw_h|_{0,\Gamma-\Sigma} + Ch^2|(\nabla.u)w_h|_0;$$

from which we deduce that : $|u\nabla w_h|_{0,\Omega} < Ch$, $|(u.n)^{1/2}w_h|_{0,\Gamma-\Sigma} \le Ch^{3/2}$ and $|(\nabla.u)^{1/2}w_h|_0 \le Ch^{3/2}$, which in turn gives

$$|w_h|_{0,\Omega} = |\phi_h - \psi_h|_{0,\Omega} \le Ch^{\frac{3}{2}} \qquad (39)$$

from which the result follows since

$$|\phi - \phi_h| \le Ch^{\frac{3}{2}} + |\psi_h - \phi| \le C'h^{\frac{3}{2}} \qquad (40)$$

Remark:
If $\nabla.u$ is not positive the proof uses the following argument:
Let $\psi = e^{-k(x)}\varphi(x)$ where $k(x)$ is any smooth function. Then with $v = e^{k(x)}u$, ψ is a solution of

$$\nabla.(v\psi) = f,$$

but $\nabla.v = e^k(\nabla.u + u\nabla k)$, so one can choose k so has to have $\nabla.v > 0$.

2.5. Scheme 4 (Upwinding by discontinuity) :

If ϕ is approximated by polynomial functions which are discontinuous on the sides (faces) of the elements, one can introduce an upwinding scheme via the integrals on the sides (faces), using the values of ϕ_h at the right or the left of the sides depending on the flow direction.

Lesaint [148] has introduced one of the first methods of this type and has given error estimates for Friedrich systems. In the case of equation (10) let us search ϕ_h in the space of functions which are piecewise polynomials but not necessarily continuous on a triangulation $\cup T_j$ of Ω :

$$W_h = \{w_h : w_h|_T \in P^k\} \qquad (41)$$

We approximate (10) by

$$\int_T w_h \nabla.(u\phi_h) - \int_{\partial-T} u.n[\phi_h]w_h = \int_T fw_h \quad \forall w_h \in W_h \quad \forall T \qquad (42)$$

where ϕ_h is taken in W_h, n is the external normal to ∂T, $[\phi_h]$ denotes the jump of ϕ_h across ∂T from upstream of ∂T :

$$[\phi_h](x) = lim_{\epsilon \to 0+}(\phi_h(x + \epsilon u(x)) - \phi_h(x - \epsilon u(x))), \quad \forall x \in \partial T \qquad (43)$$

and $\partial^- T$ is the part of ∂T where u.n < 0.If T intersects Γ^- the convention is that $\phi_h = \phi_\Gamma$ on the other side of Γ^- i.e. $\phi_h(x - \epsilon u(x)) = \phi_\Gamma(x)$ when $x \in \Gamma^-$.

Remark 1:
If $\nabla.u = 0$ and $k = 0$ (ϕ_h piecewise constant) (42) becomes

$$-\int_{\partial^- T} u.n[\phi_h] = \int_T f \quad \forall T \quad i.e.$$

$$\phi_h|_T = [\sum_j \phi_h|_{T_j} \int_{\partial^- T \cap \partial T_j} |u.n| + \int_T f]/\int_{\partial^- T} |u.n| \tag{44}$$

where T_j is the triangle upstream of T which shares $\partial^- T$. In one dimension with $u = 1$ and $diam(T) = h$ it reduces to $\phi_i = \phi_{i-1} + hf$, for all i.

One can prove that the scheme is positive if all the streamlines cut Σ :

$$f \geq 0, \phi_\Gamma \geq 0 \Rightarrow \phi_h \geq 0 \tag{45}$$

Remark 2 :

If the solution ϕ is continuous, we expect that ϕ_h will tend to a continuous function also in (42) ; so we have added to the standard weak formulation a term which is small (the integral on $\partial^- T$).

Proposition 5 :

If $\nabla.u = 0$, if all the streamlines cut Σ and if h is small, (42) defines ϕ_h in a unique manner

Proof :

We suppose for simplicity that $\phi_\Gamma \in W_h$.

Let ϵ_h be the difference between the 2 solutions. Let T be a triangle such that $\partial^- T$ is in Σ. Then (42) with $w_h = \epsilon_h|_T$ on T, 0 otherwise, gives

$$\int_T u\nabla\epsilon_h\epsilon_h - \int_{\partial^- T} u.n[\epsilon_h]\epsilon_h = 0 \tag{46}$$

$$= -\frac{1}{2}\int_{\partial^- T} u.n\epsilon_h^2 + \int_{\partial T - \partial^- T} u.n\frac{\epsilon_h^2}{2} + \int_{\partial^- T} u.n\epsilon_h|_T \epsilon_h|_{T^-}$$

where T^- is the triangle upstream of T tangent to ∂T^-. If ∂T^- is in Σ then (46) becomes:

$$\int_{\partial T - \partial^- T} u.n\frac{\epsilon_h^2}{2} = 0.$$

So $\epsilon_h = 0$ on $\partial T - \partial^- T$. Thus step by step we can cover the domain Ω.

Convergence :

Johnson -Pitkäranta [127] have shown that this scheme is of $0(h^{k+1/2})$ for the error in $L^2 - norm$ and $0(h^{k+1})$ if the triangulation is uniform ; this result was extended by Ritcher [24] who showed that the scheme is of $0(h^{k+1})$ if the triangulation is uniform in the direction of the flow only. The proofs are rather involved so we give only the general argument used in [127] and we shall only show the following partial result:

Proposition 6

If $k = 0$ (piecewise constant elements), φ is regular, $\nabla.u = 0$, and $f=0$ then

$$||\varphi - \varphi_h||_u \leq C\sqrt{h}$$

where

$$||a||_u = \left(\sum_T \int_{\partial T} |u.n|(a^+ - a^-)^2 + \int_\Gamma |u.n|a^2\right)^{1/2}$$

As in [127], we define $\Gamma^+ = \Gamma - \Sigma$ and we introduce the following notation:

$$< a, b >_h = \sum_T \int_{\partial T - \Gamma \cap \partial T} |u.n|a.b,$$

$$< a, b >_S = \int_S |u.n|a.b.$$

With this notation the discrete problem can be written in two ways:

$$< \varphi_h^+ - \varphi_h^-, w_h^- >_h + < \varphi_h, w_h >_\Sigma = < \varphi_\Gamma, w_h >_\Sigma \quad \forall w_h \in W_h$$

$$- < w_h^+ - w_h^-, \varphi_h^+ >_h + < \varphi_h, w_h >_{\Gamma^+} = < \varphi_\Gamma, w_h >_\Sigma \quad \forall w_h \in W_h$$

So if we add the two forms we see that φ_h is also a solution of:

$$< \varphi_h^+ - \varphi_h^-, w_h^- >_h - < w_h^+ - w_h^-, \varphi_h^+ >_h + < \varphi_h, w_h >_\Gamma$$
$$= 2 < \varphi_\Gamma, w_h >_\Sigma \quad \forall w_h \in W_h$$

By taking $w_h = \varphi_h$ we find that the scheme is stable:

$$||\varphi_h||_u \leq C|\varphi_\Gamma|_{0,\Gamma}.$$

The reader can establish the consistency of the scheme in the same way and obtain :

$$< \varphi^+ - \varphi^-, w_h^- >_h - < w_h^+ - w_h^-, \varphi^- >_h + < \varphi, w_h >_\Gamma = 2 < \varphi_\Gamma, w_h >_\Sigma$$

So for all ψ_h in W_h

$$||\varphi_h - \varphi||_u \leq ||\psi_h - \varphi||_u$$

If ψ_h is a piecewise constant interpolation of φ one has $|\varphi - \psi_h| = ||\varphi||_1 O(h)$, so

$$\left(\sum \int_{\partial T} |u.n|O(h)^2\right)^{\frac{1}{2}} \leq C\sqrt{h}$$

and the result follows.

2.6 Scheme 5 (Upwinding by cell)

The drawback of the previous method is that there are many degrees of freedom in discontinuous finite elements (in 3D there are on the average 5 times more tetrahedra, at least, than nodes).

To rectify this, Baba-Tabata [7] associated to all conforming P^1 function a function P^o on cells centered around the vertices and covering Ω.

To each node q^i we associate a cell σ^i defined by the medians not originating from q^i, of the triangles having q^i as one of the vertices.

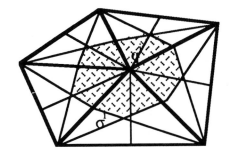

Figure 3.6: *The cell σ^i associated to vertex q^i* .

Let ϕ_h be a function $P^1 - piecewise$ on a triangulation and continuous. To ϕ_h we associate a function ϕ'_h, piecewise P^0 on the cells σ^i by the formula

$$\phi'_h|_{\sigma^i} = \frac{1}{|\sigma^i|} \int_{\sigma^i} \phi_h(x) \tag{47}$$

Conversely, knowing ψ'_h, P^0 on the σ^i, one could associate to it a function ψ_h in P^1 by

$$\psi_h(q^i) = \psi'_h|_{\sigma^i} \tag{48}$$

Then we apply the scheme 4 to ϕ'_h but we store in the computer memory only ϕ_h :

This means that our problem is now to find ϕ_h, P^1 continuous on a triangulation, equal to $\phi_{\Gamma h}$ on Σ and satisfying (in the case where $\nabla . u = 0$)

$$- \int_{\partial - \sigma^i} u.n[\phi'_h] = \int_{\sigma^i} f \quad \forall i \tag{49}$$

where

$$\phi'_{h|\sigma^i} = \frac{1}{|\sigma^i|} \int_{\sigma^i} \phi_h(x) \tag{50}$$

To show that this scheme is well posed (i.e. that ϕ_h is uniquely defined) one must show that (50) has a unique solution when ϕ'_h is known because we have already shown that (49) has a unique solution when all the streamlines cut Σ. This is difficult in general but easy to show if all the triangles are equilateral. Then an elementary computation of integrals gives

$$\int_{\sigma_i} \phi_h = \frac{7}{9} \int_{\Omega} \phi_h w^i + \frac{2}{9} |\sigma^i| \phi^i \qquad (51)$$

and multiplying by ϕ_i and summing, we see that

$$\sum_i \phi^i \int_{\sigma_i} \phi_h = 0 \Rightarrow \frac{7}{9} \int_{\Omega} \phi_h^2 + \frac{2}{9} \sum |\sigma^i|(\phi^i)^2 = 0 \Rightarrow \phi_h = 0 \qquad (52)$$

2.7 Comparison of the schemes

So far we have presented three classes of methods for the stationary convection diffusion equation. It is not possible to rank the methods since each has its advantages and drawbacks: for instance, upwinding by characteristics gives good results but is conceptually more difficult and harder to program while SUPG is conceptually simple but perhaps too diffusive if an adjustment parameter is not added. Experience has shown to the author that personal preferences (such as scientific background or previous experience with one method) are the determining factors in the choice of a particular method; beyond that we can only make the following broad remarks:

1. Upwinding by the discretisation of total derivatives has been very successful with the incompressible Navier-Stokes equations where conservativity is not critical.

2. Streamline diffusion (SUPG) has been very successful in the engineering environment because it is conceptually simple.

3. Upwinding by discontinuity has been remarkably successful with the compressible Euler equations because it is the right framework to generalize finite difference upwinding schemes into finite elements. (It has even received the name of Finite Volumes in this context when used with quadrangles).

3. CONVECTION-DIFFUSION 1 :

3.1. Generalities :
In this section, we·consider the equation

$$\phi_{,t} + \nabla.(u\phi) - \nu\Delta\phi = f \text{ in } \Omega \times]0, T[\qquad (53)$$

$$\phi(x, 0) = \phi^o(x) \text{ in } \Omega; \qquad (54)$$

$$\phi|_\Gamma = \phi_\Gamma \qquad (55)$$

and knowing that in general for fluid applications ν is small, we search for schemes which work even with $\nu = 0$. Evidently, if $\nu = 0$, (55) should be relaxed to

$$\phi|_\Sigma = \phi_\Gamma \qquad (56)$$

If $\nu = 0$, (53), (54), (55) is a particular case of stationary convection (eq (10)) for the variables $\{x, t\}$ on $\Omega \times]0, T[$ with velocity $\{u, 1\}$.

Thus one could simply add a diffusion term to the previous schemes and get a satisfying theory. But if we use the "cylindrical" structure of $\Omega \times]0, T[$, we can devise two new methods, one implicit in time without upwinding and the other semi-implicit but unconditionally stable. We begin by recalling an existence and uniqueness result for (53) and a good test problem to compare the numerical schemes: the rotating hump.

Test Problem.

Consider the case where Ω is the square $]-1, 1[^2$ and where u is a velocity field of rotation around the origin

$$u(x, t) = \{y, -x\}.$$

We take $\phi_\Gamma = 0$, $f = 0$ and

$$\phi^0(x) = e^{(|x - x_0|^2 - r^2)^{-2}} \quad \forall x \text{ such that } |x - x^0| < r,$$

$$\phi^0(x) = 0 \text{ elsewhere.}$$

When $\nu = 0$, ϕ^0 is convected by u so if we look at $t \to \{x, y, \phi(x, y, t)\}$ we will see the region $\{x, y : \phi(x, y) \neq 0\}$ rotate without deformation; the solution is periodic in time and we can overlap ϕ and ϕ^0 after a time corresponding to one turn. If $\nu \neq 0$ the phenomenon is similar but the hump flattens due to diffusion; we may also get boundary layers near the boundary

3.2. Some theoretical results on the convection-diffusion equation.

Proposition 7:

Let Ω be an open bounded set with boundary Γ Lipschitz; we denote by n its exterior normal ; the system

$$\phi_{,t} + u\nabla\phi + a\phi = f \text{ in } Q = \Omega \times]0, T[\qquad (57)$$

$$\phi(x, 0) = \phi^0(x) \quad \forall x \in \Omega \qquad (58)$$

$$\phi(x, t) = g(x, t) \quad \forall (x, t) \in \Sigma = ((x, t) : u(x, t).n(x) < 0) \qquad (59)$$

has a unique solution in $C^0(0, T; L^2(\Omega))$ when $\phi^0 \in L^2(\Omega)$, $g \in C^0(0, T; L^2(\Gamma))$ and $a, u \in L^\infty(Q)$, Lipschitz in x, $f \in L^2(Q)$.

Proof :

We complete the proof by constructing the solution. Let $X(\tau)$ the solution
of

$$\frac{d}{d\tau}X(\tau) = u(X(\tau), \tau) \text{ if } X(\tau) \in \Omega \tag{60}$$

$$= 0 \text{ otherwise}$$

with the boundary condition

$$X(t) = x. \tag{61}$$

If u is the velocity of the fluid, then X is the trajectory of the fluid particle that
passes x at time t. With $u \in L^\infty(Q)$ problem (60)-(61) has a unique solution.
As X depends on the parameters x, t, we denote the solution $X(x, t; \tau)$; it is
also the "characteristic" of the hyperbolic equation (57).

We remark that :

$$\frac{d}{d\tau}\phi(X(x, t; \tau), \tau), \tau)|_{\tau=t} = \phi_{,t}(x, t) + u\nabla\phi(x, t) \tag{62}$$

So (57) can be rewritten as :

$$\phi_{,\tau} + a\phi = f \tag{63}$$

and integrating

$$\phi(x, t) = e^{-\int_0^t a(X(\tau), \tau)d\tau}[\lambda + \int_0^t f(X(\sigma), \sigma)e^{\int_0^\sigma a(X(\tau), \tau)d\tau}d\sigma] \tag{64}$$

we determine λ from (58) and (59).

If $X(x, t; 0) \in \Gamma$, then

$$\lambda = g(X(x, t; 0), \tau(x)) \tag{65}$$

where $\tau(x)$ is the time $(< t)$ for $X(x, t; \tau)$ to reach Γ.

If $X(x, t; 0) \in \Omega$, then

$$\lambda = \phi^o(X(x, t; 0)) \tag{66}$$

Corollary $(a = \nabla.u)$:

With the hypothesis of the proposition 7 and if $u \in L^2(0, T; H(div, \Omega)) \cap$
$L^\infty(Q)$, Lipschitz in x, problem (53), (58), (59) with $\nu = 0$ has a unique solution
in $C^0(0, T; L^2(\Omega))$.

-(we recall that $H(div, \Omega) = \{u \in L^2(\Omega)^n : \nabla.u \in L^2(\Omega)\}$).

Proposition 8 :

Let Ω be a bounded open set in R^n with Γ Lipschitz . Then problem

$$\phi_{,t} + \nabla.(u\phi) - \nabla.(\kappa\nabla\phi) = f \text{ in } Q = \Omega\times]0,T[\qquad (67)$$

$$\phi(x,0) = \phi^0(x) \text{ in } \Omega \qquad (68)$$

$$\phi = g \text{ on } \Gamma\times]0,T[\qquad (69)$$

has a unique solution in $L^2(0,T;H^1(\Omega))$ if $\kappa_{i,j}$, $u_j \in L^\infty(\Omega)$, $u_{i,j} \in L^\infty(\Omega)$, $f \in L^2(Q)$, $\phi^o \in L^2(\Omega)$, $g \in L^2(0,T;H^{1/2}(\Gamma))$, and if there exists a > 0 such that

$$\kappa_{ij}Z_iZ_j \geq a|Z|^2 \quad \forall Z \in R^n. \qquad (70)$$

Proof :
See Ladyzhenskaya [138].

3.3. Approximation of the Convection-Diffusion equation by discretisation first in time then in space.

In this paragraph we shall analyze some schemes obtained by using finite difference methods to discretise $\partial\phi/\partial t$ and the usual variational methods for the remainder of the equation. Let us consider equation (67) with (70) and assume, for simplicity from now on and throughout the chapter that $\kappa_{ij} = \nu\delta_{ij}, \nu > 0$, $u \in L^\infty(Q)$ and

$$\nabla.u = 0 \text{ in } Q \quad u.n = 0 \text{ on } \Gamma\times]0,T[\qquad (71)$$

$$\phi = 0 \text{ on } \Gamma\times]0,T[\qquad (72)$$

As before, Ω is also assumed regular.

The reader can extend the results with no difficulty to the case $u.n \neq 0$, with the help of paragraph 2.

As usual, we divide $]0,T[$ into equal intervals of length k and denote by $\phi^n(x)$ an approximation of $\phi(x,nk)$.

3.3.1. Implicit Euler scheme :

We search $\phi_h^{n+1} \in H_{oh}$, the space of polynomial functions of degree p on a triangulation of Ω, continuous and zero on Γ such that for all $w_h \in H_{0h}$ we have

$$\frac{1}{k}(\phi_h^{n+1} - \phi_h^n, w_h) + (u^{n+1}\nabla\phi_h^{n+1}, w_h) + \nu(\nabla\phi_h^{n+1}, \nabla w_h) = (f^{n+1}, w_h) \quad (73)$$

This problem has a unique solution because this is an $N \times N$ linear system, N being the dimension of H_{0h} :

$$(A + kB)\Phi^{n+1} = kF + I\Phi \qquad (74)$$

where $A_{ij} = (w^i, w^j) + \nu k(\nabla w^i, \nabla w^j)$, $B_{ij} = (u^{n+1}\nabla w^i, w^j)$, $F_i = (f, w^i)$, $I = (w^i, w^j)$ and where $\{w^i\}$ is a basis of H_{0h} and ϕ_i the coefficients of ϕ_h on this basis. This system has a unique solution because the kernel of $A + kB$ is empty :

$$0 = \psi^T(A + kB)\psi = \psi^T A\psi \Rightarrow \psi = 0. \qquad (75)$$

Proposition 9 :
If $\phi \in L^2(0, T; H^{p+1}(\Omega))$ and $\phi_{,t} \in L^2(0, T; H^p(\Omega))$, we have

$$(|\phi_h^n - \phi(nk, .)|_0^2 + \nu k|\nabla(\phi_h^n - \phi(nk, .))|_0^2)^{\frac{1}{2}} \le C(h^p + k) \qquad (76)$$

where p is the degree of the polynomial approximation for ϕ_h^n.

Proof :
a) *Estimate of the error in time*
Let ϕ^{n+1} be the solution of the problem discretised in time only :

$$\frac{1}{k}(\phi^{n+1} - \phi^n) + u^{n+1}\nabla\phi^{n+1} - \nu\Delta\phi^{n+1} = f^{n+1}; \phi^{n+1}|_\Gamma = 0; \phi^o = 0 \qquad (77)$$

the error $\epsilon^n(x) = \phi^n(x) - \phi(nk, x)$ satisfies $\epsilon^n|_\Gamma = 0$, $\epsilon^o = 0$ and

$$\frac{1}{k}(\epsilon^{n+1} - \epsilon^n) + u^{n+1}\nabla\epsilon^{n+1} - \nu\Delta\epsilon^{n+1} = k\int_0^1 (1 - \theta)\frac{\partial^2\phi}{\partial t^2}(x, (n + \theta)k)d\theta \qquad (78)$$

where the right hand side is the result of a Taylor expansion of $\phi(nk + k, x)$. Multiplying (78) by ϵ^{n+1} and integrating on Ω makes the convection term disappear and we deduce that

$$||\epsilon^{n+1}||_\nu \equiv (|\epsilon^{n+1}|_0^2 + k\nu|\nabla\epsilon^{n+1}|_0^2)^{\frac{1}{2}} \le k^2 T|\phi_{tt}''|_{0,\Omega\times]0,T[} + ||\epsilon^n||_\nu. \qquad (79)$$

So we have, for all n:

$$||\epsilon^n||_\nu \le k\sqrt{T}|\phi_{,tt}''|_{0,Q} \qquad (80)$$

b) *Estimation of the error in space*
Let $\xi^n = \phi_h^n - \phi^n$; by subtracting (77) from (73) we see

$$\frac{1}{k}(\xi^{n+1} - \xi^n, w_h) + (u^{n+1}\nabla\xi^{n+1}, w_h) + \nu(\nabla\xi^{n+1}, \nabla w_h) = 0 \qquad (81)$$

Let ψ_h^n be the interpolation of ϕ^n ; then $\xi^n = \xi_h^n + \psi_h^n - \phi^n$ and ξ_h^n satisfies (we take $w_h = \xi_h^{n+1}$)

$$\|\xi_h^{n+1}\|_\nu^2 \leq |\xi_h^n|_o |\xi_h^{n+1}|_o - (\psi_h^{n+1} - \phi^{n+1} - \psi_h^n + \phi^n, \xi_h^{n+1}) - \qquad (82)$$

$$-k(u^{n+1}\nabla(\psi_h^{n+1} - \phi^{n+1}), \xi_h^{n+1}) - \nu k(\nabla(\psi_h^{n+1} - \phi^{n+1}), \nabla\xi_h^{n+1})$$

so

$$\|\xi_h^{n+1}\|_\nu \leq \|\xi_h^n\|_\nu + Ck[h^p\|\phi'_{,t}\|_{p,\Omega} + h^p(\|u\|_{\infty,Q} + \nu)\|\phi\|_{p+1,\Omega}] \qquad (83)$$

because $|.|_o \leq \|.\|_\nu$, $\psi_h^{n+1} - \psi_h^n$ is the interpolation of $\phi^{n+1} - \phi^n$ which is an approximation of $k\phi'_{,t}$.

Finally we obtain (76) by noting that

$$\|\phi_h^n - \phi(nk)\|_\nu \leq \|\xi_h^n\|_\nu + \|\psi_h^n - \phi^n\|_\nu + \|\phi(nk) - \phi^n\|_\nu \qquad (84)$$

Comments :
We note that the error estimate (76) is valid even with $\nu = 0$. Thus, we do not need any upwinding in space. This is because the Euler scheme itself is upwinded in time (it is not symmetric in n and $n+1$). The constant c in (76) depends on $\|\phi\|_{p+1,\Omega}$ and that if ν is small (67) has a boundary layer and $\|\phi\|_{p+1,\Omega}$ tends to infinity (if ϕ_Γ is arbitrary) as $\nu^{-p+1/2}$. This requires the modification of ϕ_Γ or the imposition of: $h << \nu^{1-1/2p}$.

A third method consists of replacing ν by ν_h in such a way that $h << \nu^{1-1/2p}$ is always satisfied. This is the artificial viscosity method.

3.3.2. Leap frog scheme :
The previous scheme requires the solution of a non-symmetric matrix for each iteration. This is a scheme which does not require that kind of operation but which works only when k is $O(h)$.

$$\frac{1}{2k}(\phi_h^{n+1} - \phi_h^{n-1}, w_h) + (u^n\nabla\phi_h^n, w_h) + \frac{\nu}{2}(\nabla[\phi_h^{n+1} + \phi_h^{n-1}], \nabla w_h) \qquad (85)$$

$$= (f^{n+1}, w_h) \quad \forall w_h \in H_{0h}, \quad \phi_h^{n+1} \in H_{0h}$$

To start (85), we could use the previous scheme.

Proposition 10 :
The scheme (85) is marginally stable , i.e. there exists a C such that

$$|\phi_h^n|_o \leq C|f|_{0,Q}(1 - C|u|_\infty \frac{k}{h})^{-\frac{1}{2}} \qquad (86)$$

for all k such that

$$k < \frac{h}{C|u|_\infty}. \qquad (87)$$

Proof in the case where u is independent of t and f = 0.

We use the energy method (Ritchmeyer-Morton [197], Saiac [211] for example).

Let

$$S_n = |\phi_h^{n+1}|_0^2 + |\phi_h^n|_0^2 + 2k(u\nabla\phi_h^n, \phi_h^{n+1}) \qquad (88)$$

then

$$S_n - S_{n-1} = |\phi_h^{n+1}|_0^2 - |\phi_h^{n-1}|_0^2 + 2k(u\nabla\phi_h^n, \phi_h^{n+1} + \phi_h^{n-1}) \qquad (89)$$

but from (85), with $w_h = \phi_h^{n+1} + \phi_h^{n-1}$ and if $f = 0$

$$|\phi_h^{n+1}|_0^2 - |\phi_h^{n-1}|_0^2 + 2k(u\nabla\phi_h^n, \phi_h^{n+1} + \phi_h^{n-1}) + k\nu|\nabla(\phi_h^{n+1} + \phi_h^{n-1})|_0^2 = 0; \quad (90)$$

so $S_n \leq S_{n-1} \leq \dots \leq S_o$. On the other hand, using an inverse inequality

$$(u\nabla\phi_h^n, \phi_h^{n+1}) \geq -\frac{C}{h}|u|_\infty|\phi_h^{n+1}|_0|\phi_h^n|_0 \geq -|u|_\infty\frac{C}{2h}(|\phi_h^{n+1}|_0^2 + |\phi_h^n|_0^2) \qquad (91)$$

we get

$$S_n \geq (1 - C\frac{k|u|_\infty}{h})[|\phi_h^{n+1}|_0^2 + |\phi_h^n|_0^2] \qquad (92)$$

Convergence :

As for the Euler scheme, one can show using (86) that (85) is $0(h^p + k^2)$ for the L^2 norm error .

3.3.3. Adams-Bashforth scheme :

It is important to make the dissipation terms implicit as we have done for $\Delta\phi$ because the leap-frog scheme is only marginally stable (cf. Richtmeyer-Morton [197]). In the same way, if $\Sigma \neq \oslash$ ($u.n \neq 0$), it is necessary to make implicit the integral on $\Gamma-\Sigma$. For this reason, we consider the Adams-Bashforth scheme of order 3 which is explicit when we use mass lumping and which has a better stability than the leap-frog scheme.

$$\frac{1}{k}(\phi_h^{n+1} - \phi_h^n, w_h) = \frac{23}{12}b(\phi_h^n, w_h) - \frac{16}{12}b(\phi_h^{n-1}, w_h) + \frac{5}{12}b(\phi_h^{n-2}, w_h)$$

$\forall w_h \in H_{0h}$ where

$$b(\phi_h, w_h) = -[(u\nabla\phi_h, w_h) + \nu(\nabla\phi_h, \nabla w_h) - (f, w_h)]$$

The stability and convergence of the scheme can be analysed as in §3.3.5.

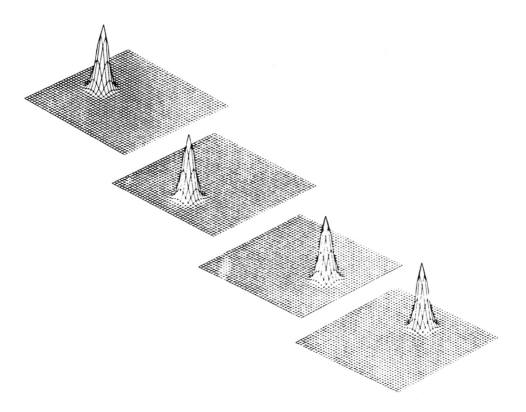

Figure 3.7 : *Solution of the rotating hump test problem by discretisation of the total derivative with Q2 conforming quadrilateral elements and a 9 points Gauss quadrature formula The time step is approximately h/u. The surface $\{x, y, \varphi(x, y)\}$ is plotted at 4 time instants. Pointwise errors are below à 2% . (Computed by par M. Hasbani).*

3.3.4 The $\theta-$ schemes :

In a general way, let A and B be two operators and the equation in time be

$$u_{,t} + Au + Bu = f; u(0) = u^0$$

We consider a scheme with three steps

$$\frac{1}{k\theta}(u^{n+\theta} - u^n) + Au^{n+\theta} + Bu^n = f^{n+\theta}$$

$$\frac{1}{(1 - 2\theta)k}(u^{n+1-\theta} - u^{n+\theta}) + Au^{n+\theta} + Bu^{n+1-\theta} = f^{n+1-\theta}$$

$$\frac{1}{k\theta}(u^{n+1} - u^{n+1-\theta}) + Au^{n+1} + Bu^{n+1-\theta} = f^{n+1}$$

An analysis of this scheme in the finite element context can be found in Glowinski [96]. One can show easily that with $A = \alpha C$, $B = (1 - \alpha)C$, where C is an matrix $N \times N$ with strictly positive eigenvalues, then the scheme is unconditionally stable and of order 2 in k if $\alpha = 1/2$ and $\theta = 1 - \sqrt{2}/2$.

In the case of the convection-diffusion equation, we could take $A = -\alpha\nu\Delta$ and $B = u\nabla - (1 - \alpha)\nu\Delta$ ($\alpha = 1$ being admissible). Chorin [53], Beale-Majda [17] have studied methods of this type where $A = -\nu\Delta$ and steps 1 and 3 are carried out by a Monte-Carlo method, and where step 2 is integrated by a finite element method, a finite difference method or by the method of characteristics. We will use these for the Navier-Stokes equations.

3.3.5. Adaptation of the finite difference techniques :

All the methods presented up to now in this section have been largely studied on regular grids in a finite difference framework. The same techniques can be used in a finite element context to estimate stability and errors, with the following restrictions :
- constant coefficients (u and ν constants),
- uniform triangulation
- influence of boundary conditions is difficult to take into account

The analysis of finite difference schemes is based on the following fundamental property:

Stability + consistency $\Rightarrow$ convergence.

To give an example, let us study the Crank-Nicolson scheme for (67) in the case where u is constant :

$$\frac{1}{k}(\phi_h^{n+1} - \phi_h^n, w_h) + \frac{1}{2}(u\nabla(\phi_h^{n+1} + \phi_h^n), w_h) + \frac{\nu}{2}(\nabla(\phi_h^{n+1} + \phi_h^n), \nabla w_h)$$

$$= (f^{n+\frac{1}{2}}, w_h) \quad \forall w_h \in H_{0h}$$

Although the L^2 stability is simple to prove by taking $w_h = \phi_h^{n+1} + \phi_h^n$, we will consider the other methods at our disposal.

By choosing a basis $\{w^i\}$ of H_{0h} we could write explicitly the linear system corresponding to the case $f = 0$ and $\phi_h|_\Gamma = 0$.

$$B(\Phi^{n+1} - \Phi^n) + \frac{k}{2}A(\Phi^{n+1} + \Phi^n) = 0$$

where $B_{ij} = (w^i, w^j)$ $A_{ij} = \nu(\nabla w^i, \nabla w^j) + (u\nabla w^i, w^j)$

If we know the eigenvalues and eigenvectors of A in the metric B, i.e. the solutions $\{\lambda_i, \psi^i\}$ of $\lambda B\psi = A\psi$, then by decomposing Φ^n on this basis, we get

$$(B\Phi)_i^{n+1} = (B\Phi)_i^n \frac{(1 - \frac{k}{2}\lambda_i)}{(1 + \frac{k}{2}\lambda_i)}$$

By asking the amplification factor $F = (1-k\,\lambda_i/2)/(1 + k\,\lambda_i/2)$ to be smaller than 1 , we can deduce an interval of stability of the method.

Evidently, the smallest real part of the eigenvalues is not known but could be numerically determined in the beginning of the calculation (as in Maday et al. [164] for the Stokes problem).

On the other hand, when H_{0h} is constructed with $P^1 - continuous$ functions, the Crank-Nicolson scheme on a uniform triangulation is identical to the following finite difference scheme (exercise) :

$$\frac{h^2}{12}[6\phi_{i,j}^{n+1} + \phi_{i+1,j+1}^{n+1} + \phi_{i+1,j}^{n+1} + \phi_{i,j-1}^{n+1} + \phi_{i-1,j-1}^{n+1} + \phi_{i-1,j}^{n+1} + \phi_{i,j+1}^{n+1}]$$

$$+\frac{kh}{12}[(\phi_{i+1,j+1}^{n+1} - \phi_{i-1,j-1}^{n+1})(u_1 + u_2) + (2u_1 - u_2)(\phi_{i+1,j}^{n+1} - \phi_{i-1,j}^{n+1})$$

$$+(2u_2 - u_1)(\phi_{i,j+1}^{n+1} - \phi_{i,j-1}^{n+1})] + \nu k[4\phi_{i,j}^{n+1} - \phi_{i+1,j}^{n+1} - \phi_{i,j+1}^{n+1} - \phi_{i-1,j}^{n+1} - \phi_{i,j-1}^{n+1}]$$

$$= \text{ idem in } \phi^n \text{ by changing } k \text{ to } -k.$$

If there exists solutions of the form :

$$\phi_{lm}^n = \psi^n e^{2i\pi(lx+my)}$$

they have to satisfy

$$\psi^{n+1}[\frac{h^2}{6}[3 + cos((l + m)h) + cos(lh) + cos(mh)]$$

$$+i\frac{kh}{6}[sin((l + m)h)(u_1 + u_2) + sin(lh)(2u_1 - u_2) + sin(mh)$$

$$(2u_2 - u_1)] + 2\nu k[2 - cos(lh) - cos(mh)]]$$

$$= \psi^n[\text{ same factor but } k \to -k]$$

So we have a formula for the amplification factor.

Finally, it is easy to see that the above finite difference scheme is consistent to order 2. So we have a presumption of convergence of $0(h^2 + k^2)$ for the method on a general triangulation .

4.CONVECTION DIFFUSION 2.

In this section we analyze schemes for the convection-diffusion equation (53)-(55) which still converge when $\nu = 0$ without generating oscillations.

This classification is somewhat arbitrary because the previous schemes can be made to work when $\nu = 0$ or when φ is irregular. But we have put in this section schemes which have been generalized to nonlinear equations (Navier-Stokes and Euler equations for example). Let us list the desirable properties for a scheme to work on nonsmooth functions ϕ :

    - convergence in L^∞ norm,
    - positivity: $\varphi > 0 \Rightarrow \varphi_h > 0$,
    - convergence to the stationary solution when $t \to \infty$,
    - localization of the solution if $\nu = 0$ (that is to say that the solution should not depend upon whatever is downstream of the characteristic when $\nu = 0$).

4.1. Discretisation of the total derivative:

4.1.1 Discretisation in time.

We have seen that if $X(x, t; \tau)$ denotes the solution of

$$\frac{dX}{d\tau}(\tau) = u(X(\tau), \tau); \quad X(t) = x \tag{93}$$

then

$$\phi_{,t} + u\nabla\phi = \frac{\partial}{\partial\tau}\phi(X(x, t; \tau), \tau)|_{\tau=t} \tag{94}$$

Thus, taking into account the fact that $X(x, (n + 1)k; (n + 1)k) = x$, we can write:

$$(\phi_{,t} + u\nabla\phi)^{n+1} \cong \frac{1}{k}[\phi^{n+1}(x) - \phi^n(X^n(x))] \tag{95}$$

where $X^n(x)$ is an approximation of $X(x, (n + 1)k; nk)$.

We shall denote by X_1^n an approximation $0(k^2)$ of $X^n(x)$ and by X_2^n an approximation $0(k^3)$ (the differences between the indices of X and the exponents of k are due to the fact that X^n is an approximation of X obtained by an integration over a time k ; thus a scheme $0(k^\alpha)$ gives a precision $0(k^{\alpha+1})$).

For example

$$X_1^n(x) = x - u^n(x)k \text{ (Euler scheme for (93))} \tag{96}$$

$$X_2^n(x) = x - u^{n+\frac{1}{2}}(x - u^n(x)\frac{k}{2})k \text{ (Second order Runge-Kutta)} \tag{97}$$

modified near the boundary so as to get $X_i^n(\Omega) \subset \Omega$. To obtain this inclusion one can use (96) or (97) inside the elements so that one passes from x to $X^n(x)$ by a broken line rather than a straight line (see (18)(19)).

This yields two schemes for (53) :

$$\frac{1}{k}(\phi^{n+1} - \phi^n o X_1^n) - \nu\Delta\phi^{n+1} = f^{n+1} \tag{98}$$

$$\frac{1}{k}(\phi^{n+1} - \phi^n o X_2^n) - \frac{\nu}{2}\Delta(\phi^{n+1} + \phi^n) = f^{n+\frac{1}{2}} \tag{99}$$

Lemma 1 :

If u is regular and if $X_i^n(\Omega) \subset \Omega$, the schemes (98) and (99) are L^2−stable and converge in $0(k)$ and $0(k^2)$ respectively.

Proof :

Let us show consider (98). We multiply by ϕ^{n+1} :

$$|\phi^{n+1}|^2 + \nu k|\nabla\phi^{n+1}|^2 \le (|f^{n+1}|k + |\phi^n o X_1|)|\phi^{n+1}| \tag{100}$$

But the map $x \rightarrow X$ preserves the volume when u^n is solenoidal ($\nabla.u = 0$). So from (96) :

$$|\phi^n o X_1^n|_0^2 = \int_{X_1^n(\Omega)} \phi^n(y)^2 det[\nabla X_1^n]^{-1} dy \le |\phi^n|_{0,\Omega}^2(1 + ck^2) \tag{101}$$

Hence ϕ^n verifies

$$\|\phi^n\|_\nu \le c[|f|_{0,Q} + |\phi^o|_{0,\Omega}] \tag{102}$$

To get an error estimate one proceeds as in the beginning of the proof of proposition 9 by using (95).

Remark :

$X_i^n(\Omega) \subset \Omega$ is necessary because $u.n = 0$. Otherwise one only needs $X_i^n(\Omega) \cap \partial\Omega \subset \Sigma$.

4.1.2 Approximation in space.

Now if we use the previous schemes to approximate the total derivative (scheme (98) of order 1 , scheme (99) of order 2) and if we discretise in space by a conforming polynomial finite element we obtain a family of methods for which no additional upwinding is necessary and for which the linear systems are symmetric and time independent .

Take for example the case of (98) :

$$\int_\Omega \phi_h^{n+1} w_h + k\nu \int_\Omega \nabla\phi_h^{n+1}\nabla w_h = k\int_\Omega f^{n+1} w_h + \int_\Omega \phi_h^n(X_1^n(x))w_h(x)$$

$$\forall w_h \in H_{0h} \quad \phi_h^{n+1} \in H_{0h} \tag{103}$$

where H_{0h} is the space of continuous polynomial approximation of order 1 on a triangulation of Ω, and zero on the boundaries.

Proposition 11 :
If $X_1(\Omega) \subset \Omega$, the scheme (103) is $L^2(\Omega)$ stable even if $\nu = 0$.

Proof :
One simply replaces w_h by ϕ_h^{n+1} in (103) and derives upper bounds :

$$|\phi_h^{n+1}|_0^2 \leq \int_\Omega |\phi_h^{n+1}|^2 + k \int_\Omega \nu \nabla \phi_h^{n+1} \nabla \phi_h^{n+1} \tag{104}$$

$$= k \int_\Omega f^{n+1} \phi_h^{n+1} + \int_\Omega \phi_h^n (X_1^n(x)) \phi_h^{n+1}(x)$$

$$\leq (k|f^{n+1}|_0 + |\phi_h^n o X_1^n(.)|_0)|\phi_h^{n+1}|_0$$

$$\leq (|\phi_h^n|_0(1 + \frac{c}{2}k^2) + k|f^{n+1}|_0)|\phi_h^{n+1}|_0$$

The last inequality is a consequence of(101).
Finally by induction one obtains

$$|\phi_h^n|_{0,\Omega} \leq (1 + \frac{c}{2}k^2)^n(|\phi_h^o|_{0,\Omega} + \sum k|f^n|_{0,\Omega}) \tag{105}$$

Remark :
By the same technique similar estimates can be found for (105) but the norms on ϕ_h^n et ϕ_h^o will be $||.||_\nu$ (cf. (79)).

Proposition 12 .
If H_{0h} is a P^1 conforming approximation of $H_0^1(\Omega)$ then the $L^2(\Omega)$ norm of the error between ϕ_h^n solution of (103) and ϕ^n solution of (98) is $0(h^2/k + h)$. Thus the scheme is $0(h^2/k + k + h)$.

Proof :
One subtracts (98) from (103) to obtain an equation for the projected error:

$$\epsilon_h^{n+1} = \phi_h^{n+1} - \Pi_h \phi^{n+1}, \tag{106}$$

where $\Pi_h \phi^{n+1}$ is an interpolation in H_{0h} of ϕ^{n+1}. One gets

$$\int_\Omega \epsilon_h^{n+1} w_h + k\nu \int_\Omega \nabla \epsilon_h^{n+1} \nabla w_h - \int_\Omega \epsilon_h^n o X_1^n w_h = \tag{107}$$

$$\int_\Omega (\phi^{n+1} - \Pi_h \phi^{n+1}) w_h + \nu k \int_\Omega \nabla(\phi^{n+1} - \Pi_h \phi^{n+1}) \nabla w_h$$

$$- \int_\Omega (\phi^n - \Pi_h \phi^n) o X_1^n w_h$$

From (107), with $w_h = \epsilon_h^{n+1}$ we obtain

$$||\epsilon_h^{n+1}||_\nu^2 \leq (||\epsilon_h^n||_\nu + ||\phi^{n+1} - \Pi_h\phi^{n+1}||_\nu + |\phi^n - \Pi_h\phi^n|_0)||\epsilon_h^{n+1}||_\nu$$

therefore

$$||\epsilon_h^{n+1}||_\nu \leq ||\epsilon_h^n||_\nu + C(h^2 + \nu k h)$$

Remark :
By comparing with (103), we see that ϵ_h and ϕ_h are solutions of the same problem but for ϵ_h, f is replaced by :

$$\frac{1}{k}(\phi^{n+1} - \Pi_h\phi^{n+1}) - \nu\Delta_h(\phi^{n+1} - \Pi_h\phi^{n+1}) - \frac{1}{k}(\phi^n - \Pi_h\phi^n)oX_1^n,$$

where Δ_h is an approximation of Δ. We can bound independently the first and the last terms. By working a little harder (Douglas-Russell [69]) one can shown that the error is, in fact, $0(h + k + \min(h^2/k, h))$.

Proposition 13 :
With the second order scheme in time (99) and a similar approximation in space one can build schemes $0(h^2 + k^2 + \min(h^3/k, h^2))$ with respect to the L^2 norm:

$$(\phi_h^{n+1}, w_h) + \frac{\nu k}{2}(\nabla(\phi_h^{n+1} + \phi^n), \nabla w_h) = (\phi_h^n oX_2^n, w_h) + k(f^{n+\frac{1}{2}}, w_h) \quad (108)$$

$$\forall w_h \in H_{0h}; \phi_h^{n+1} \in H_{0h}$$

where H_{0h} is a P^2 conforming approximation of $H_0^1(\Omega)$.

Proof :
The proof is left as an exercise.

The case $\nu = 0$:
We notice that (103) becomes

$$\int \phi_h^{n+1} w_h = \int_\Omega \phi_h^n oX_2^n w_h + k \int_\Omega f^{n+1} w_h \quad \forall w_h \in H_{0h} \quad (109)$$

$$\phi_h^{n+1} \in H_{0h}$$

That is to say

$$\phi_h^{n+1} = \Pi_h(\phi_h^n oX_1^n) + k\Pi_h f^{n+1} \quad (110)$$

where Π_h is a L^2 projection operator in W_{0h}. Scheme (108) becomes :

$$\phi_h^{n+1} = \Pi_h(\phi_h^n o X_2^n) + \frac{k}{2}\Pi_h(f^{n+1} + f^n) \tag{111}$$

If $f = 0$ the only difference between the schemes are in the integration formula for the characteristics. Notice also that the numerical diffusion comes from the L^2 projection at each time step. Thus it is better to use a precise integration scheme for the characteristics and use larger time steps. Experience shows that $k \approx 1.5h/u$ is a good choice.

Notice that when ν and f are zero one solves

$$\phi_{,t} + u\nabla\phi = 0 \quad \phi(x,0) = \phi^0(x) \tag{112}$$

-(since we have assumed $u.n = 0$, no other boundary condition is needed).

Since $\nabla.u = 0$, we deduce from (112) that (conservativity)

$$\int_\Omega \phi(t,x) = \int_\Omega \phi^0(x) \quad \forall t \tag{113}$$

On the other hand, from (109), with $w_h = 1$

$$\int_\Omega \phi_h^{n+1}(x) = \int_\Omega \phi_h^n o X_1^n = \int_{X_1^n(\Omega)} \phi_h^n(y)det|\nabla X_1^n|^{-1}dy \tag{114}$$

So if $det|\nabla X_1^n| = 1$ (which requires $X_1^n(\Omega) = \Omega$), one has

$$\int_\Omega \phi_h^{n+1}(x)dx = \int_\Omega \phi_h^n(y)dy = \int_\Omega \phi^0(x)dx \tag{115}$$

We say then that the scheme is conservative. It is an important property in practice.

4.1.3. Numerical implementation problems :

Two points need to be discussed further.
- How to compute $X^n(x)$,
- How to compute I^n :

$$I^n = \int_\Omega \phi_h o X_h^n w_h \tag{116}$$

Computation of (116) :
As in the stationary case one uses a quadrature formula:

$$I^n \cong \sum \omega^k \phi_h(X_h(\xi^k))w_h(\xi^k) \tag{117}$$

For example with P^1 elements one can take
  a) $\{\xi^k\}$ = the middles of the sides, $\omega^k = \sigma_k/3$ in 2D, $\sigma_k/4$ in 3D, where σ_k is the area (volume) of the elements which contain ξ^k.

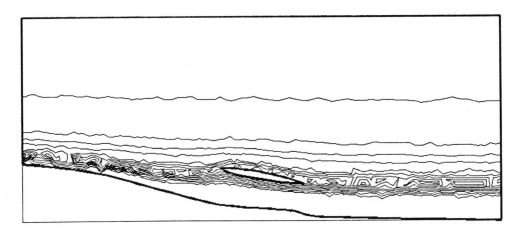

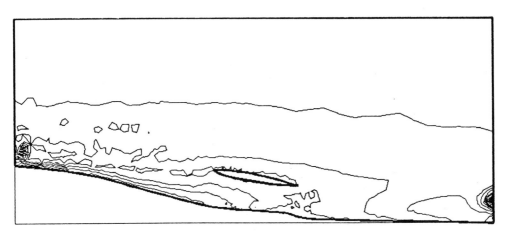

Figure 3.8 : *Numerical simulation of a stationary convection equation by characteristics and P1 conforming elements (top) and comparison with the leap-frog +P1 element scheme (bottom); in the second case the stationary solution is computed as the limit of the transient problem. (Courtesy of J.H. Saiac).*

b) The 3 (4 in 3D) point quadrature formula (Zienkiewicz [241], Stroud [224]) or any other more sophisticated formula ; but experiments show that quadrature formula with negative weights ω^k should not be chosen. Numerical tests with a 4 points quadrature formulae can be found in Bercovier et al. [27].

Finally, another method (referred as *dual* because it seems that the basis functions are convected forward) is found by introducing the following change of variable:

$$\int_\Omega \phi_h o X_h w_h = \int_{X_h(\Omega)} \phi_h(y) w_h(X_h^{-1}(y)) det|\nabla X_h^{-1}| dy \qquad (118)$$

Then the quadrature formula are used on the new integral. If $\nabla.u$ is zero one can even take

$$I \cong \sum \omega^k \phi_h(\xi^k) w_h(X_h^{-1}(\xi^k)) w_h^k \tag{119}$$

It is easy to check that this method, (109), (119), is conservative while (109), (117) is not. Numerical tests for this method can be found in Benque et al. [20].

The stability and error estimate when a quadrature formula is used is an important open problem ; it seems that they have a bad effect in the zones where u is small (Suli [226], Morton et al.[177]).

Computation of $X^n(x)$:
Formula (96) and (97) can be used directly. However it should be pointed out that in order to apply (117), (same problem with (119)) one needs to know the number l of the element such that

$$X^n(\xi^k) \in T_l \tag{120}$$

This problem is far from being simple. A good method is to store all the numbers of the neighboring (by a side) elements of each element and compute the intersections $\{\xi^k, X^n(\xi^k)\}$ with all the edges (faces in 3D) between ξ^k and $X^n(\xi^k)$; but then one can immediately improve the scheme for X_h by updating u with its local value on each element when the next intersection is searched; then a similar computation for (18)-(19) is made:
Let $\{u_i\}$ be such that $u = \sum_i u_i q^i$, $\sum u_i = 0$, where $\{q^i\}$ are the vertices of the triangle (tetrahedron):
Find ρ such that

$$\lambda_i' = \lambda_i + \rho u_i \Rightarrow \prod_i \lambda_i' = 0, \quad \lambda_i' \ge 0. \tag{121}$$

This is done by trial and error; we assume that it is λ_m which is zero, so:

$$\rho = -\frac{\lambda_m}{\mu_m}, \tag{122}$$

and we check that $\lambda_i \ge 0$, $\forall i$. If it is not so we change m until it works.

Most of the work goes into the determination of the u_i. Notice that it may be practically difficult to find which is the next triangle to cross when $X^n(\xi^k)$ is a vertex, for example. This requires careful programming.

When $|u - u_h|_\infty$ is $0(h^p)$ the scheme is $0(h^p)$. If u_h is piecewise constant one must check that $u_h.n$ is continuous across the sides (faces) of the elements (when $\nabla.u = 0$) otherwise (121) may not have solutions other than $\rho = 0$. If $u_h = \nabla \times \psi_h$ and ψ_h is P^1 and continuous then $u_h.n$ is continuous.

4.2. The Lax-Wendroff/ Taylor-Galerkin scheme[25]

Consider again the convection equation

$$\phi_{,t} + \nabla.(u\phi) = f \quad in \quad \Omega \times]0, T[\tag{123}$$

For simplicity assume that $u.n|_\Gamma = 0$ and that u and f are independent of t. A Taylor expansion in time of ϕ gives:

$$\phi^{n+1} = \phi^n + k\phi_{,t}^n + \frac{k^2}{2}\phi_{,tt}^n + 0(k^3) \tag{124}$$

If one computes $\phi_{,t}$ and $\phi_{,tt}$ from (123), one finds

$$\phi^{n+1} = \phi^n + k[f - \nabla.(u\phi^n)] - \frac{k^2}{2}\nabla.(u[f - \nabla.(u\phi^n)]) + 0(k^3) \tag{125}$$

or again

$$\phi^{n+1} = \phi^n + k[f - \nabla.(u\phi^n)] + \frac{k^2}{2}[-\nabla.(uf) + \nabla.[u\nabla.(u\phi^n)]] + 0(k^3) \tag{126}$$

This is the scheme of Lax-Wendroff [144]. In the finite element world this scheme is known as the Taylor-Galerkin method (Donea [68]). Note that the last term is a numerical diffusion $0(k)$ in the direction u^n because it is the tensor $u^n \otimes u^n$.

Let us discretise (126) with H_h, the space of P^1 continuous function on a triangulation of Ω :

$$(\phi_h^{n+1}, w_h) = ((\phi_h^n, w_h) + k(f - \nabla.(u\phi_h^n), w_h) \tag{127}$$

$$-\frac{k^2}{2}(\nabla.(uf), w_h) - \frac{k^2}{2}(u\nabla w_h, \nabla.(u\phi_h^n)) \quad \forall w_h \in H_h$$

The scheme should be $O(h^2 + k^2)$ in the L^2 norm but it has a CFL stability condition (Courant-Friedrischs-Lewy) (see Angrand-Dervieux [2] for a result of this type on an $O(h)$-regular triangulation for scheme (127) with mass lumping)

$$k < C\frac{h}{|u|_\infty} \tag{128}$$

An implicit version can also be obtain by changing n into $n+1$ and k into $-k$ in (126)

$$(\phi_h^{n+1}, w_h) + k(\nabla.(u\phi_h^{n+1}), w_h) + \frac{k^2}{2}(\nabla.(u\phi_h^{n+1}), u\nabla w_h) \tag{129}$$

$$= k(f, w_h) + (\phi_h^n, w_h) + \frac{k^2}{2}(\nabla.(uf), w_h), \quad \forall w_h \in H_h$$

In the particular case when $\nabla.u = 0$ and $u.n|_\Gamma = 0$, we have the following result :

Proposition 14 :

If u and f are independent of t and $\nabla.u = 0$, $u.n_{|\Gamma} = 0$, then scheme (129) has a unique solution which satisfies

$$|\phi_h^n|_0 \leq |\phi_h^o|_0 + T|f|_0 + \frac{T}{2}k|\nabla.(uf)|_0 \text{ (stability)} \qquad (130)$$

$$|\phi_h^n - \phi(nk)|_0 \leq C(h^2 + k^2) \text{ (convergence)} \qquad (131)$$

Proof :

The symmetric part of the linear system yielded by (129) multiplied left and right by ψ_h gives $|\psi_h|_o^2 + k^2/2 \, |u\nabla\psi_h|^2$ which is always positive ; thus the linear systems being square, they have one and only one solution.

By taking $w_h = \phi_h^{n+1}$ in (129), one finds

$$|\phi_h^{n+1}|_0^2 + \frac{k^2}{2}|u\nabla\phi_h^{n+1}|^2 \leq |\phi_h^n|_0|\phi_h^{n+1}|_0 + k|f|_0|\phi^{n+1}|_0 + \frac{k^2}{2}|\nabla.(uf)|_0|\phi_h^{n+1}|_0$$

$$(132)$$

thus if one divides by $|\phi_h^{n+1}|_o$ and adds all the inequalities, (130) is found.

To obtain (131) one should subtract (126) from (129) :

$$(\epsilon^{n+1}, w_h) + \frac{k^2}{2}(u\nabla\epsilon^{n+1}, u\nabla w_h) + k(u\nabla\epsilon^{n+1}, w_h) = (\epsilon^n + 0(k^3), w_h) \quad (133)$$

where $\epsilon = \phi_h - \phi$. Let $\epsilon_h = \phi_h - \psi_h$ where ψ_h is the projection of ϕ in H_h with the norm $||.||_k = (|.|_0^2 + k^2/2|u\nabla.|_0^2)|^{1/2}$. Then $\epsilon_h = \epsilon + 0(h^2)$ and from (133), we get

$$||\epsilon_h^{n+1}||_k \leq ||\epsilon_h^n|| + C(kh^2 + k^3) \qquad (134)$$

The result follows.

Remarks :

1. The previous results can be extended without difficulty to the case $\nu \neq 0$ with the scheme:

$$(\phi_h^{n+1}, w_h) + k(\nabla.(u\phi_h^{n+1}), w_h) + \frac{k^2}{2}(\nabla.(u\phi_h^{n+1}), u\nabla w_h) + \nu(\nabla\phi_h^{n+1}, \nabla w_h)$$

$$(135)$$

$$= k(f, w_h) + (\phi_h^n, w_h) + \frac{k^2}{2}(\nabla.(uf), w_h), \quad \forall w_h \in H_{oh}$$

Higher order schemes :

Donea also report that very good higher order schemes can obtained easily by pushing the Taylor expansion further; for instance a third order scheme would be ($f = 0$):

$$\phi^{n+1} - [\phi^n - k\nabla.(u\phi^n) + \frac{k^2}{2}\nabla.[u\nabla.(u\phi^n)] + \frac{k^2}{6}\nabla.[u\nabla.(u(\phi^{n+1} - \phi^n))]] = 0$$

4.3. The streamline upwinding method (SUPG).

The streamline upwinding method studied in §2.4 can be applied to (123) without distinction between t and x but it would then yield very large linear systems. But there are other ways to introduce streamline diffusion in a time dependent convection-diffusion equation.

The simplest (Hughes [116]) is to do it in space only; so consider

$$(\phi_h^{n+1}, w_h + \tau\nabla.(uw_h)) + \frac{k}{2}(\nabla.(u[\phi_h^n + \phi_h^{n+1}]), w_h + \tau\nabla.(uw_h))$$

$$+\frac{k\nu}{2}(\nabla(\phi_h^{n+1} + \phi_h^n), \nabla w_h) - \frac{k\nu}{2}\sum_l \int_{T_l}(\Delta(\phi_h^{n+1} + \phi_h^n)\tau\nabla.(uw_h))$$

$$= k(f^{n+\frac{1}{2}}, w_h + \tau\nabla.(uw_h)) + (\phi_h^n, w_h + \tau\nabla.(uw_h)), \quad \forall w_h \in H_{oh}$$

where T_l is an element of the triangulation and u is evaluated at time $(n+1/2)k$ if it is time dependant; τ is a parameter which should be of order h but has the dimension of a time.

With first order elements, $\Delta\phi_h = 0$ and it was noticed by Tezduyar [227] that τ could be chosen so as to get symmetric linear systems when $\nabla.u = 0$; an elementary computation shows that the right choice is $\tau = k/2$. Thus in that case the method is quite competitive, even though the matrix of the linear system has to be rebuilt at each time step when u is time dependant.

The error analysis of Johnson [125] suggests the use of elements discontinuous in time, continuous in space and a mixture upwinding by discontinuity in time and streamline upwinding in space.

To this end space-time is triangulated with prisms. Let $Q^n = \Omega\times]nk, (n+1)k[$, let W_o^n be the space of functions $in\{x, t\}$ which are zero on $\Gamma\times]nk, (n+1)k[$ continuous and piecewise affine in x and in t separately on a triangulation by prisms of Q^n ;

We search ϕ_h^n with $\phi_h^n - \phi_{\Gamma h} \in W_{oh}^n$ solution of

$$\int_{Q^n} [\phi_{h,t}^n + \nabla(u\phi_h^n)][w_h + h(w_{h,t} + \nabla(uw_h)]dxdt + \int_{Q^n} \nu\nabla\phi_h^n.\nabla w_h dxdt \quad (136)$$

$$+ \int_\Omega \phi_h^n(x, (n-1)k+0)w_h(x, (n-1)k)dx = \int_{Q^n} f(w_h + h(w_{h,t} + \nabla.(uw_h)))dxdt$$

$$+ \int_\Omega \phi_h^{n-1}(x,(n-1)k-0)w_h(x,(n-1)k)dx, \quad \forall w_h \in W_{oh}^n$$

Note that if N is the number of vertices in the triangulation of Q^n, equation (136) is an $N \times N$ linear system, positive definite but non symmetric.

One can show (Johnson et al. [127]) the following :

$$(\int_0^T |\phi_h^n - \phi|_{0,\Omega}^2 dt)^{\frac{1}{2}} \le C(h^{\frac{3}{2}} + k^{\frac{3}{2}})\|\phi\|_{H^2(Q)}. \qquad (137)$$

4.4. Upwinding by discontinuity on cells:

The method of §2.5 can be extended to the nonstationary case but there is an approximation problem for $\Delta\phi$ when ϕ_h is discontinuous in space. One way is to use the upwinding by cells as in §2.6 ; since the scheme will be 0(h) one can use a piecewise constant approximation in time, which is similar to a discretisation of (53) (see (49) for the notation) by

$$(\phi_h^{n+1}, w^i) - \frac{k}{|\sigma^i|} \int_{\partial-\sigma^i} u.n[\int_{\sigma^i} \phi_h^{n+1}dx]d\gamma + \nu k(\nabla\phi_h^{n+1}\nabla w^i) =$$

$$= (\phi_h^n, w^i) + k(f, w^i), \quad \forall i, \quad \phi_h^{n+1} - \phi_{\Gamma h} \in H_{oh}$$

where w^i is the continuous piecewise affine function associated with the ith vertex of the triangulation of Ω.

This scheme is used in Dervieux et al [64][65] for the Euler equations (see Chapter 6).

Chapter 4

The stokes problem

1. POSITION OF THE PROBLEM

The generalized Stokes problem is to find $u(x) \in R^n$ and $p(x) \in R$ such that

$$\alpha u - \nu \Delta u + \nabla p = f \quad \nabla . u = 0 \text{ in } \Omega, \tag{1}$$

$$u = u_\Gamma \text{ on } \Gamma = \partial \Omega. \tag{1'}$$

where α, ν are given positive constants and f is a function from Ω into R^n.

It can come from at least two sources :

1) an approximation of the fluid mechanics equations such as that seen in Chapter 1, that is when the Reynolds number is small (microscopic flow, for example), then $\alpha = 0$, and in general $f = 0$;

2) a time discretisation of the Navier-Stokes equations. Then α is the inverse of the time step size and f an approximation of $-u\nabla u$.

In this chapter, we shall deal with some finite element approximations of (1). More details regarding the properties of existence, unicity and regularity can be found in Ladyzhenskaya [138], Lions [153] and Temam [228] and regarding the numerical approximations in Girault-Raviart [93], Thomasset [229], Girault-Raviart [93], Glowinski [95], Hughes [116] .

Test Problem :
The most classic test problem is the cavity problem :

The domain Ω is a square $]0,1[^2$, $\alpha = 0, \nu = 1, f = 0$ and the boundary condition:

$u = 0$ on all the boundary except the upper boundary $]0,1[\times\{1\}$ where $u = (1.,0)$.

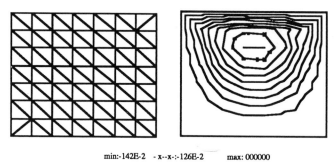

min:-142E-2   - x--x-:-126E-2   max: 000000

Figure 4.1 : *Stokes flow in a cavity*

This problem does not have much physical signifigance but it is easy to compare with solutions computed by finite differences. We note that the solution has singularities at the two corners where u is discontinuous (the solution is not in H^1).We can regularize this test problem by considering on the upper boundary $]0,1[\times\{1\}$:

$$u = x(1-x)$$

2. FUNCTIONAL SETTING.

Let n be the dimension of the physical space ($\Omega \subset R^n$) and

$$J(\Omega) = \{u \in H^1(\Omega)^n : \quad \nabla.u = 0\} \tag{2}$$

$$J_o(\Omega) = \{u \in J(\Omega) : \quad u|_\Gamma = 0\} \tag{3}$$

Let us put

$$(a,b) = \int_\Omega a_i b_i \quad a,b \in L^2(\Omega)^n \tag{4}$$

$$(A,B) = \int_\Omega A_{ij} B_{ij} \quad A,B \in L^2(\Omega)^{n \times n} \tag{5}$$

and consider the problem

$$\alpha(u,v) + \nu(\nabla u, \nabla v) = (f,v) \quad \forall v \in J_o(\Omega) \tag{6a}$$

$$u - u'_\Gamma \in J_o(\Omega) \tag{6b}$$

where u'_Γ is an extension in $J(\Omega)$ of u_Γ.

Lemma 1
If $u_\Gamma \in H^{1/2}(\Gamma)^n$ and if $\int_\Gamma u_\Gamma.n = 0$ then there exists an extension $u'_\Gamma \in J(\Omega)$ of u_Γ.

Remarks
1. For notational convenience we assume that u_Γ always admits an extension with zero divergence and we identify u_Γ and u'_Γ.
2. Taking account of this condition and without loss of generality, we arrive by translation $(v = u - u'_\Gamma)$ at the case $u_\Gamma = 0$.

Theorem 1
If $f \in L^2(\Omega)^n$ and $u'_\Gamma \in J(\Omega)$, problem (6) has a unique solution.

Proof :
$J_o(\Omega)$ is a non empty closed subspace of $H^1(\Omega)^n$ and the bilinear form

$$\{u, v\} \rightarrow \alpha(u, v) + \nu(\nabla u, \nabla v)$$

is $H_0^1(\Omega)$ elliptic. With the hypothesis $v \rightarrow (f, v)$ continuous, the theorem is a direct consequence of the Lax-Milgram theorem.

Remark
We also have a variational principle : (6) is equivalent to

$$\min_{u - u'_\Gamma \in J_o(\Omega)} \frac{1}{2}\alpha(u, u) + \frac{1}{2}\nu(\nabla u, \nabla u) - (f, u) \tag{7}$$

Theorem 2
If the solution of (6) is C^2 then it satisfies (1). Conversely, under the hypothesis of theorem 1, if $\{u, p\} \in H^1(\Omega)^n \times L^2(\Omega)$ is a solution of (1) then u is a solution of (6).

Proof :
We apply Green's formula to (6) :

$$\int_\Omega (\alpha u - \nu \Delta u).v = \int_\Omega fv \quad \forall v \in J_o(\Omega) \tag{8}$$

Now we use the following theorem (Cf Girault-Raviart [92] for example)

$$\int_\Omega g.v = 0 \quad \forall v \in L^2(\Omega) \text{ with } \nabla.v = 0 \quad v.n|_\Gamma = 0 \quad \Rightarrow \tag{9}$$

$$\text{There exists } q \in L^2(\Omega) \text{ such that } g = \nabla q \tag{10}$$

- We recall that $(\nabla p, v) = 0$ for all $v \in J_o(\Omega)$.

So (8) implies the existence of p $\in L^2(\Omega)$ such that

$$\alpha u - \nu\Delta u = f - \nabla p \tag{11}$$

and the result follows. Conversely, by multiplying (1) with v $\in J_o(\Omega)$ and integrating we obtain (6) since $v_\Gamma = 0$:

$$-\int_\Omega \Delta uv = \int_\Omega \nabla u\nabla v \text{ and } -\int_\Omega p\nabla.v = \int_\Omega v\nabla p$$

Remark

1. It is not necessary that u be C^2 for p to exist, but then the proof is based on a more abstract version of (9)(10). Indeed it is enough to notice that the linear map :

$$L(v) : v \rightarrow (f, v) - \alpha(u, v) - \nu(\nabla u, \nabla v)$$

is zero on $J_0(\Omega)$ because the space orthogonal to $J(\Omega)$ is the space of gradients, thus there exists a q in $L^2(\Omega)$ such that $L(v) = (\nabla q, v)$ for all v in $H_0^1(\Omega)^n$; hence (1) is true in the distribution sense.

2. Problem (1) can also be directly studied in the form of a saddle point problem (Cf. Girault-Raviart [92] for example),

$$\alpha(u, v) + \nu(\nabla u, \nabla v) - (\nabla.v, p) = (f, v) \quad \forall v \in H_0^1(\Omega)^n$$

$$(\nabla.u, q) = 0 \quad \forall q \in L^2(\Omega)/R$$

We shall use later this formulation in the error analysis of the methods.

3. The uniqueness of the pressure, p (up to a constant) is not given by theorem 1 but it can be shown by studying the saddle point problem.

3. DISCRETISATION

Let $\{J_h\}_h$ be a sequence in a finite dimensional space, $J_{oh} = J_h \cap H_0^1(\Omega)^n$, such that

$$\forall v \in J_o(\Omega) \text{ there exists } v_h \in J_{oh} \text{ such that } \|v_h - v\|_1 \rightarrow 0 \text{ when } h \rightarrow 0 \quad (12)$$

We consider the approximated problem :

Find u_h such that

$$\alpha(u_h, v_h) + \nu(\nabla u_h, \nabla v_h) = (f, v_h) \quad \forall v_h \in J_{oh}; \quad u_h - u'_{\Gamma_h} \in J_{oh}. \tag{13}$$

where u'_{Γ_h} is an approximation of u'_{Γ} in J_h.

Theorem 3
If J_{oh} is non empty the problem (13) has a unique solution.

Proof
Let us consider the problem

$$\min_{u_h - u'_{\Gamma_h} \in J_{oh}} \frac{1}{2}\{\alpha(u_h, u_h) + \nu(\nabla u_h \nabla u_h) - (f, u_h)\} \tag{14}$$

Since J_{oh} is of finite dimension N, (14) is an optimization of a strictly quadratic function in N variables ; thus it admits a unique solution. By writing the first order optimality conditions for (14), we find (13).

A counter example .
Let Ω be a quadrilateral and T_h be the triangulation of 4 triangles formed with the diagonals of Ω. Let

$$J_{0h} = \{v_h \in C^0(\Omega): \quad v_h|_{T_k} \in P^1 \quad v_h|_{\Gamma_h} = 0,$$

$$(\nabla.v_h, q) = 0 \quad \forall q \text{ continuous, piecewise affine on } T_h\}$$

There are only two degrees of freedom for v_h (its values at the center) but there are 4 constraints of which 3 are independent, so J_{0h} is reduced to $\{0,0\}$.

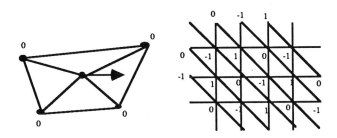

Figure 4.2 :*values of the velocity at the vertices of the first counterexample and the pressure for the second.*

Remarks
1. One might think that the problem with the above counter example stems from the fact that there are not enough vertices. Although J_{oh} would not, in general, be empty when there are more vertices, this approximation is not good, as we shall see, because there are too many independent constraints. For example, let Ω be a square triangulated into $2(N-1)^2$ triangles formed by the straight lines parallel to the x and y axis and the lines at $45°$ (figure 4.2). We have also $2N^2$ constraints and $2(N^2 - 4N) \cong 2N^2$ degrees of freedom when N is large.

2. The pressure cannot be unique in that case. Indeed with J_{0h} defined as above there exists $\{p_h\}$ piecewise affine continuous such that

$$(\nabla.v_h, p_h) = 0 \quad \forall v_h \quad P^1 \text{ continuous on } T_h$$

These are the functions p_h which have values alternatively 0, 1 and -1 on the vertices of each triangle (each triangle having exactly these three values at the three vertices.) Then $(\nabla.v_h, p_h)$ is zero because

$$(\nabla.v_h, p_h) = \sum_{T_i} (\nabla.v_h)|_{T_i} \left(\frac{1}{3} \sum_{j=1,2,3} p_h(q^{ij}) \right)$$

So the constraints in J_{oh} are not independent and the numerical pressure associated with the solution of (14) (the Lagrangian multiplier of the constraints) cannot be unique.

To solve numerically (13) (or (14)) we can try to construct a basis of J_{oh}. Let $\{v^i\}_i^N$ be a basis of J_{oh}. By writing u_h on this basis,

$$u_h(x) = \sum_{1..N} u^i v^i(x) + u'_{\Gamma_h}(x), \tag{15}$$

it is easy to see by putting (15) in (13) with $v_h = v^i$ that (13) reduces to a linear system

$$AU = G \tag{16}$$

with

$$A_{ij} = \alpha(v^i, v^j) + \nu(\nabla v^i, \nabla v^j), \tag{17}$$

$$G_j = (f, v^j) - \alpha(u'_{\Gamma_h}, v^j) - \nu(\nabla u'_{\Gamma_h}, \nabla v^j). \tag{18}$$

However there are two problems

1) It is not easy to find an internal approximation of $J(\Omega)$; that is in general we don't have $J_{oh} \subset J_0(\Omega)$, which leads to considerable complications in the study of convergence.

2) Even if we succeed in constructing a non empty J_{oh} which satisfies (12) and for which we could show convergence, the construction of the basis $\{v^i\}$ is in general difficult or unfeasible which makes (15)(16) inapplicable in practice.

Before solving these problems, let us give some examples of discretisation for J_{oh} in increasing order of difficulty. These examples are all convergent and feasible as we shall see later. They are all of the type

$$J_h = \{v_h \in V_h : \quad (\nabla.v_h, q)_h = 0 \quad \forall q \in Q_h\}$$

$$J_{0h} = \{v_h \in J_h : \quad v_h|_\Gamma = 0\}$$

Thus J_{0h} is determined by the choice of two spaces
  1°) V_h which approximate $H^1(\Omega)$,
  2°) Q_h which approximate $L^2(\Omega)$ or $L^2(\Omega)/R$
  3°) and the choice of a quadrature formula. $(.,.)_h$.

As usual, $\{T_k\}$ denotes a triangulation of Ω, Ω_h the union of T_k, $\{q^j\}_{j=1..Ns}$ the vertices of the triangulation, $\lambda_j^k(x)$ the $j-th$ barycentric coordinate of x with respect to vertices $\{q^{k_j}\}_{j=1.,n+1}$ of the element T_k. We denote respectively by

  Ns, the number of vertices,
  Ne, the number of elements,
  Nb, the number of vertices on Γ,
  Na, the number of sides of the triangulation, and
  Nf, the number of faces of the tetrahedra in 3D.

P^m denotes the space of polynomials (of degree $\leq$ m) in n variables. We recall Euler's geometric identity :

$$Ne - Na + Ns = 1 \text{ in 2D} , Ne - Nf + Na - Ns = -1 \text{ in 3D}$$

3.1. P1 bubble/P1 element (Arnold-Brezzi-Fortin [5])

Let $\mu^k(x)$ be the bubble function associated with an element T_k defined by :

$$\mu^k(x) = \prod_{j=1..n+1} \lambda_j^k(x) \text{ on } T_k \text{ and 0 otherwise}$$

The function μ^k is zero outside of T_k and on the boundary and positive in the interior of T_k. Let $w^i(x)$ be a function defined by

$$w^i(x) = \lambda_i^k(x) \text{ on all the } T_k \text{ which contain the vertex } i \text{ and 0 otherwise}$$

Let us put :

$$V_h = \{\sum_j v^j w^j(x) + \sum_k b^k \mu^k(x) : \quad \forall v^j, b^k \in R^n\} \tag{19}$$

that is the set of (continuous) functions having values in R^n and being the sum of a continuous piecewise affine function and a linear combination of bubbles. We shall remark that $dim V_h = (Ns + Ne)n$.
  Let

$$Q_h = \{\sum_j p^j w^j(x) : \quad \forall p^j \in R\} \tag{20}$$

that is a set of piecewise affine and continuous functions (usual P^1 element).
We have : dim Q_h = Ns

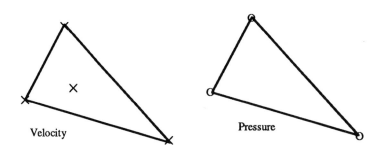

Figure 4.3:*Position of the degrees of freedom in the $P^1 - bubble/P^1$ element*

Let us put

$$J_h = \{v_h \in V_h : (\nabla.v_h, q_h) = 0 \quad \forall q_h \in Q_h\} \tag{21}$$

$$J_{oh} = \{v_h \in J_h : v_h|_{\Gamma_h} = 0\}. \tag{22}$$

It is easy to show that all the constraints (except one) in (21) are independent. In fact if N is the dimension of V_h the constraints which define J_h are of the form $BV = 0$ where V are the values of v_h at the nodes. So if we show that

$$(v_h, \nabla q_h) = 0 \quad \forall v_h \in V_h \quad \Rightarrow \quad q_h = constant$$

that proves that Ker B^T is of dimension 1 and so the ImB is of dimension $N-1$, i.e.B is of rank $N - 1$ which is to say that all the constraints are independent except one (the pressure is defined up to a constant).

Let e^i be the ith vector in the cartesian system. Let us take $b^k = \delta_{k,i}e^i$, $v^j = 0$ and v_h in (19). Then

$$0 = (v_h, \nabla q_h) = \frac{\partial q_h}{\partial x_i}|_{T_i} \frac{area(T_i)}{(n+1)} \Rightarrow \quad q_h = constant$$

So we have ($n = 2$ or 3):

$$dim J_h = (n-1)(Ns + Ne)n - Ns + 1 = Ns + nNe + 1 \tag{23}$$

We shall prove that this element leads to an error of $0(h)$ in the $H^1 norm$ for the velocity. Whereas, without knowing a basis (local in x) one cannot solve (13) and one has to use duality for the constraints. A basis, local in x, for J_h

is not known (by local in x, we mean that each basis function v^i has a support around q^i and it is zero far from q^i).

Remark

1. We can replace the bubble by a function zero on ∂T_k piecewise affine on 3 internal triangles of each triangle obtained by dividing it at its center ; we obtain the same convergence result.

2. The bubbles are very easily eliminated in practice because they are orthogonal to the P^1 -basis functions $((\nabla w^i, \nabla w^b) = 0)$. As pointed out by Lohner and Pierre, they are equivalent to an artificial dissipation $-\nabla.(c(h,x)\nabla p)$ added to (1b) where $(c(h,x)|_{T_b} = (w^b,1)^2/[(\nabla w^b, \nabla w^b)area(T_b)]$.

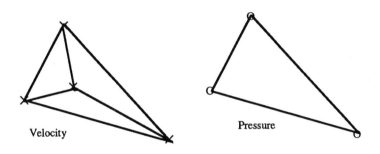

Figure 4.4:*Position of the degrees of freedom in the element* $P_{h'}^1/P^1$.

3.2. P1 iso P2/P1 element (Bercovier-Pironneau [26])

We construct a triangulation $T_{h/2}$ from the triangle T_h in the following manner :

Each element is subdivided into s sub-elements ($s = 4$ for triangles and 8 for tetrahedra) by the midpoints of the sides.

We put

$$V_h = \{v_h \in C^0(\Omega_h)^n : v_h|T_k \in (P^1)^n \quad \forall T_k \in T_{\frac{h}{2}}\} \qquad (24)$$

$$Q_h = \{q_h \in C^0(\Omega_h) : q_h|T_k \in P^1 \quad \forall T_k \in T_h\}. \qquad (25)$$

We construct J_{oh} by (21)-(22)

Knowing that the number of sides of a triangulation is equal to $Ne+Ns-1$ if n=2 (Ns+Nf-Ne-1 if n=3) we get, in 2-D :

$$dimJ_h = 2Ns + 2(Ns + Ne - 1) - Ns + 1 = 3Ns + 2Ne - 1 \qquad (26)$$

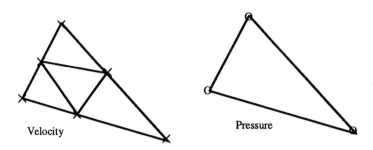

Figure 4.5 : *Degrees of freedom in the $P^1 isoP^2/P^1$ element* .

The independence of the constraints is shown in the following way :

We take $v_h \in V_h$ with 0 on all nodes (vertices and midpoints of sides) except at the middle node q^l of the side a_l, intersection of the triangles T_{k_1} and T_{k_2}. Then :

$$0 = (v_h, \nabla q_h) = v_h(q^l)[\sum_{j=1,2} \nabla q_h|_{T_{k_j}} area(T_{k_j})]$$

Let q_{l_1} and q_{l_2} be the values of q_h at the two ends of a_l; if we take $v_h(q^l) = q_{l_1} - q_{l_2}$ then the above constraint gives $q_{l_1} = q_{l_2}$. So the q_h which satisfy the constraints for all v_h are constant.

This element also gives an error of $0(h)$ for the $H^1 norm$. We know how to construct a basis with zero divergence usable for this element (Hecht [109]).

3.3. P2/P1 element (Hood-Taylor [113])

$$V_h = \{v_h \in C^0(\Omega_h)^n : \quad v_h|_{T_k} \in (P^2)^n \quad \forall k\} \tag{27}$$

$$Q_h = \{q_h \in C^0(\Omega_h) : \quad q_h|_{T_k} \in P^1 \quad \forall k\} \tag{28}$$

and of course (21)-(22).

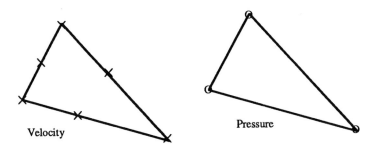

Figure 4.6 : *Position of the degrees of freedom in the* P^2/P^1 *element* .

This element is $0(h^2)$ for the H^1 error and has the same dimension (26) but it gives matrices with bandwidth larger than (24)-(25). The proof of the independence constraints is the same as above. The dimension of J_h is therefore:

$$dimJ_h = 2(Ns + Ne - 1) + 2Ns - Ns + 1 = 3Ns + 2Ne - 1$$

Construction of a zero divergence basis :

We know how to construct a basis with zero divergence usable for this element (Hecht [109]) ; figure 4.7 gives the non zero directions at the nodes of each basis function with zero divergence, as well as its support (set of points where it is non zero).

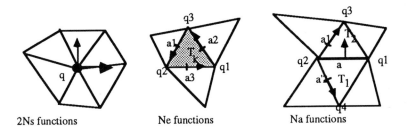

Figure 4.7: *Basis with zero divergence for the* P^2/P^1 *element* .

For the 3 families of functions, the notations are the following : a denotes a side, T a triangle and q a vertex. We have drawn the triangles on which the function is non zero and the directions of the vectors, values of these functions at the nodes, when they are non zero.

Since all constraints but one in J_h are independent ; the dimension of J_h is

$$dim J_h = dim V_h - dim Q_h = 2Ns + 2Na - Ns + 1 = 2Ns + Ne + Na$$

the last equality being a consequence of Euler's identity.

We shall define three families of vectors :

- $2Ns$ vectors, each with 0 on all the nodes (middle of the sides and vertices) except a vertex q where the vector is arbitrary .

- Ne vectors v^k, each having value 0 on all the nodes except on the 3 mid-points of the sides of a triangle where $v^k = \sum_{1..3} \lambda_j a_t^j$ with, for example :

$$\lambda_1 = 1, \quad \lambda_2 = \frac{(\nabla w^3, a_t^1)}{(\nabla w^3, a_t^2)}, \quad \lambda_3 \doteq \frac{(\nabla w^2, a_t^1)}{(\nabla w^3, a_t^3)}$$

where a_t denotes a unitary vector tangent to the side a and w^i the basis function P^2 associated to the vertex q^i.

- Na vectors v non-zero only on a side a and on a side a^1 and a^2 of the 2 triangles which contain a. If a_n denotes the unitary vector normal to a we could take, for example $v = \lambda a_n + \mu a_t^1 + \nu a_t^2$ with :

$$\lambda = -\frac{(\nabla w^1, a_n)}{(\nabla w^2, a_t)}, \quad \mu = -\frac{(\nabla w^3, a_n)}{(\nabla w^3, a_t^1)}, \quad \nu = -\frac{(\nabla w^4, a_n)}{(\nabla w^4, a_t^2)}$$

We leave it to the reader to verify that the vectors are independent and with zero divergence in the sense of (21)(27)(28).

3.4. P1 nonconforming/P0 element (Crouzeix-Raviart [60])

$$V_h = \{v_h \text{ continuous at the middle of the sides (faces) } v_h|_{T_k} \in (P^1)^n\} \quad (29)$$

$$Q_h = \{q_h : q_h|_{T_k} \in P^0\} \quad (30)$$

and (21)-(22).

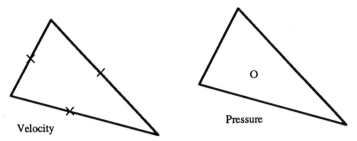

Figure 4.8:*Position of degrees of freedom for the element P^1 nonconforming $/P^0$.*

This element is $0(h)$ in the $H^1 norm$; it is nonconforming because v_h is discontinuous at the interfaces of the elements. The dimension of V_h is equal to twice the number of sides (resp. faces). We prove the independence of the constraints by taking v_h perpendicular to a side and zero at all the other nodes:

$$0 = (\nabla.v_h, q_h) = \sum \int_{side} q_h v_h.nd\gamma = \sum \text{ length(side) } (q_h|_{T_{k_1}} - q_h|_{T_{k_2}})$$

$$\Rightarrow \quad q_h = constant.$$

In R^2 we have

$$dim J_h = Ne + 2Ns - 1 \tag{31}$$

We know how to construct a local basis with zero divergence for this element. In dimension 2 we define two families of basis vectors of J_h, one with indices on the middle of the sides q^{ij} :

$$v'^{ij}(q^{kl}) = (q^i - q^j)\frac{\delta_{ik}\delta_{jl}}{|q^i - q^j|} \quad \forall q^{kl} \text{ middle of the side } \{q^k, q^l\} \notin \Gamma \tag{32}$$

and the other with indices on the internal vertices q^l

$$v'^l(q^{ij}) = \delta_{li}\frac{n(q^{ij})}{|q^i - q^j|} \quad \forall q^{ij} \text{ middle of } \{q^i q^j\} \tag{33}$$

where n is normal to $q^i q^j$ in the direct sense.

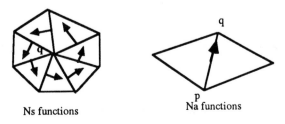

Figure 4.9: *Zero divergence basis for the P^1 nonconforming P^0 element .*

We remark that for the two families we have :

$$(\nabla.v', q_h) = \sum_k q_h|_{T_k} \int_{T_k} v'.n dx = 0 \quad \forall q_h \in Q_h \qquad (34)$$

because each integral is zero. Moreover if Na denotes the number of sides there are $Na + Ns$ basis functions in this construction ; that number is also equal to $2Ns + Ne - 1$ which is exactly the dimension of J_h (cf. (31)); so we have a basis of J_h.

In 3-D a similar construction is possible (Hecht [109]) with the middle of the faces and the vertices but the functions thus obtained are no longer independent; one must remove all the functions v'^l of a maximal tree without cycle made of edges.

3.5. P2bubble/P1 discontinuous element . (Fortin [80])

$$V_h = \{v_h \in C^0(\Omega_h)^n : v_h(x) = w_h + \sum_l b_l \mu^l(x), \quad w_h|_{T_l} \in (P^{k+1})^n, \quad b_l \in R^n\}$$
$$(35)$$
$$Q_h = \{q_h : q_h|_{T_k} \in P^k\} \ (q_h \text{ is not continuous}), \qquad (36)$$

$k = 0$ or 1. With (21)-(22) this element is $0(h^2)$ in norm H^1 if $k = 1$ and $0(h)$ if $k = 0$. Since the constraints are independent (except one as usual) we can easily deduce the dimension of J_{0h}. The construction of a basis with zero divergence for this element can also be found in [15]. We note that if we remove the bubbles and if $k = 1$ the element becomes $0(h)$.

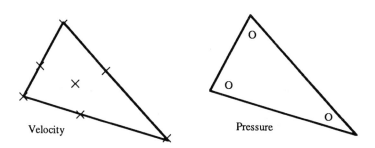

Velocity   Pressure

Figure 4.10: *Position of degrees of freedom for the element*
$P^2 bubble/P^1$ discontinuous .

Remark

One can construct without difficulty a family of elements with quadrilaterals by replacing in the formulae P^m by Q^m, the space of polynomials of degree $\leq m$ with respect to *each* of its variables.

3.6 Comparison of elements:

The list below is only a small subset of possible elements tfor the solution of the Stokes problem. We have only given the elements which are most often used. The choice of an element is based on the three usual criteria :
- memory capacity,
- precision,
- programming facility.

Even with recent developments in computer technology the memory capacity is still the determining factor and so we have given the elements in the order of increasing memory requirement in 3D for the tetrahedra ; but that could change for computers with 256 Mwords like the CRAY 2 which could handle elements which are now considered to be too costly !

4. RESOLUTION OF THE LINEAR SYSTEMS

If we know a basis of J_{oh} and of J_h it is enough to write u_h in that basis, construct A and G by (17)-(18) and solve (16). This is a simple and efficient method but relatively costly in memory because A is a big band matrix.

To optimize the memory we store only the non zero elements of A in a one dimensional array again denoted by A. I and J are integer arrays where the indices of rows and columns of non zero elements are stored.

$m = 0$
Loop on i and j
If $a_{ij} \neq 0$, *do* $m = m + 1$, $A(m) = a_{ij}$,$I(m) = i$, $J(m) = j$
end of loop .
$m_{Max} = m$.

This method is well suited particularly to the solution of the linear system

$$AU = G$$

by the conjugate gradient method (see Chapter 2) because it needs only the following operations :

$$\text{given } U \text{ calculate } V \equiv AU;$$

now this operation is done easily with array A, I and J :
To get $V(i)$ one proceeds as follows :

Initialize $V = 0$.
For $m = 1.. \ m_{Max}$ do

$$V(I(m)) = A(m) * U(J(m)) + V(I(m))$$

4.1. Resolution of a saddle point problem by the conjugate gradient method.

All the examples given for J_{oh} are of the type

$$J_{oh} = \{v_h \in V_{oh} : (\nabla.v_h, q_h) = 0 \quad \forall q_h \in Q_h\} \qquad (36)$$

where V_{oh} is the space of functions of V_h zero on the boundaries. An equivalent saddle point problem associated with (14) is

$$\min_{u_h - u'_{\Gamma_h} \in V_{oh}} \max_{p^i} \{\frac{1}{2}\alpha(u_h, u_h) + \frac{1}{2}\nu(\nabla u_h, \nabla u_h) - (f, u_h) + \sum_{1..M} p^i(\nabla.u_h, q^i)\} \qquad (37)$$

where $\{q^i\}^M$ is a basis of Q_h.

Theorem 4

When J_{oh} is given by (36) and is non empty with $\dim V_h = N$, $\dim Q_h = M$ and $V_{oh} \subset H_0^1(\Omega)^n$ (Ω is polygonal), the discretised Stokes problem (13) is equivalent to (37). In addition, (37) is equivalent to

$$\alpha(u_h, v_h) + \nu(\nabla u_h, \nabla v_h) + (\nabla p_h, v_h) = (f, v_h) \quad \forall v_h \in V_{oh} \qquad (38)$$

$$(\nabla.u_h, q_h) = 0 \quad \forall q_h \in Q_h \qquad (39)$$

$$u_h - u'_{\Gamma_h} \in V_{oh}, \quad p_h \in Q_h$$

These are the necessary and sufficient conditions for u_h to be a solution of (13) because they are the first order optimality conditions for (14).

Proof

Problem (13) is equivalent to (14) which is a problem with quadratic criteria and linear constraints. The min-max theorem applies and the conditions for the qualification of the constraints are verified (Cf. Ciarlet [56], Luenberger [162] Polak [193] for example). The q^i are the M Lagrangian multipliers with M constraints in (36). The equivalence of (37) with (38)-(39) follows from the argument that if **L** is the criteria in (37)

$$\frac{\partial \mathbf{L}}{\partial u^i} = 0, \quad \frac{\partial \mathbf{L}}{\partial p^i} = 0 \qquad (40)$$

at the optimum, where u^i are the components of u_h expressed in the basis of V_{oh}. These are the necessary and sufficient conditions for $\{u_h, p_h\}$ to be the solution of (37) because **L** is strictly convex on V_{oh} in u_h and linear in p^i .

Remarks .

1. (37)(38) can be found easily from (1); this formulation was used in engineering well before it was justified.

2. Without the inf-sup condition (79), there is no unicity of $\{p^i\}_i$.

Proposition 1
If $\{v^i\}^N$ is a basis of V_{oh} and $\{q^j\}^M$ a basis of Q_h, problem (38)-(39) is equivalent to a linear system

$$\begin{pmatrix} A & B \\ B^T & 0 \end{pmatrix} \begin{pmatrix} U \\ P \end{pmatrix} = \begin{pmatrix} G \\ 0 \end{pmatrix} \tag{41}$$

with U being the coefficient of u_h expressed on the basis $\{u^i\}$ and P being the coefficients of p_h expressed on the basis $\{q^i\}$.

$$A_{ij} = \alpha(v^i, v^j) + \nu(\nabla v^i, \nabla v^j), \quad i, j = 1..N \tag{42}$$

$$B_{ij} = (\nabla q^i, v^j) \quad i = 1..M, \quad j = 1..N \tag{43}$$

$$G_j = (f, v^j) - \alpha(u_{\Gamma_h}, v^j) - \nu(\nabla u_{\Gamma_h}, \nabla v^j) \quad j = 1..N \tag{44}$$

Proof
We note that if u_h is zero on Γ_h we have

$$(\nabla.u_h, q_h) = -(u_h, \nabla q_h) \tag{46}$$

Then we decompose u_h and p_h on their basis and take $v_h = v^j$ and $q_h = q^i$ in (38) and (39).

Proposition 2
Problem (41) is equivalent to

$$(B^t A^{-1} B) P = B^t A^{-1} G, \quad U = -A^{-1} B P + A^{-1} G \tag{47}$$

Proof
It is sufficient to eliminate U from (41) with the first equations. This can be done because A is invertible as shown in (48) :

$$U^t A U = \alpha(u_h, u_h) + \nu(\nabla u_h, \nabla u_h) = 0 \Rightarrow u_h = 0 \Rightarrow U = 0 \tag{48}$$

Algorithm 2 (solving (47) by the Conjugate Gradient Method)
    0 . *Initialisation: choose p^0 (=0 if no initial guess known). Choose $C \in R^{N \times N}$ positive definite, choose $\epsilon \ll 1$, $nMax \gg 1$; (p^n, q^n, z, g^n are in the pressure space while u^n, v are in the velocity space)*

$$\text{solve } Au^0 = G - Bp^0 \tag{49}$$

$$\text{put } g^0 = B^t u^0 \quad (= B^t A^{-1} G) \quad n = 0. \tag{50}$$

1 . *Solve*

$$Av = Bq^n \tag{51}$$

$$and \; Cz = B^t v \quad (= B^t A^{-1} B q^n) \tag{52}$$

$$set \; \rho = \frac{|g^n|_C^2}{< q^n, z >_C} \tag{53}$$

2. *Put*

$$p^{n+1} = p^n - \rho q^n \tag{54}$$

$$u^{n+1} = u^n - \rho v \; (so \; that \; Au^{n+1} = -Bp^{n+1} + G) \tag{55}$$

$$g^{n+1} = g^n + \rho z \tag{56}$$

3. IF $(|g^{n+1}| < \epsilon)or(n > nMax)$ THEN *stop* ELSE*Put*

$$\gamma = \frac{|g^{n+1}|_c^2}{|g^n|_c^2} \tag{57}$$

$$q^{n+1} = g^{n+1} + \gamma q^n \tag{58}$$

and Return to 1 with n:=n+1 .

Choice of the preconditioning.

Evidently the steps (51) and (56) are costly. We note that (51) can be decomposed into n Dirichlet sub-problems for each of the n components of u_h. This is a *fundamental* advantage of the algorithm because all the manipulation is done only on a matrix not bigger than a scalar Laplacian matrix.

(56) is also a discrete linear system. A good choice for C (Benque and al [20]) can be obtained from the Neumann problem on Ω for the operator $-\Delta$, when we use for Q_h an admissible approximation of $H^1(\Omega)$ (Cf. 3.3, 3.2 and 3.1):

$$(\nabla \phi_h, \nabla q_h) = (l, q_h) \quad \forall q_h \in Q_h/R \quad \Rightarrow \tag{59}$$

$$C_{ij} = -\Delta_{hij} \equiv (\nabla q^i, \nabla q^j) \quad \forall i,j \; (boundary \; points \; included) \; . \tag{60}$$

This choice is based on the following observation : from (1) we deduce

$$-\Delta p = \nabla . f \; in \; \Omega, \quad \frac{\partial p}{\partial n} = f.n + \nu \Delta u.n$$

So in the case $\nu \ll 1$ there underlies a Neumann problem.

This preconditioning can be improved as shown by Cahouet-Chabard [47] by taking instead of $-\Delta_h$:

$$C \equiv (\nu I_h^{-1} - \alpha \Delta_h^{-1})^{-1} \; with \; Neumann \; conditions \; on \; the \; boundary$$

where I_h is the operator associated with the matrix (w^i, w^j) where w^i is the canonical basis function associated with the vertex q^i. In fact, the operator $B^T A^{-1} B$ is a discretisation of $\nabla.(\alpha I - \nu \Delta)^{-1} \nabla$. Now if $\alpha = 0$, it is the identity and if $\nu = 0$, it is $-\Delta_h$.

4.2. Resolution of the saddle point problem by penalization

The matrix of problem (41) is not positive definite.

Rather let us consider

$$\begin{pmatrix} A & B \\ B^t & -\epsilon I \end{pmatrix} \begin{pmatrix} U \\ P \end{pmatrix} = \begin{pmatrix} G \\ 0 \end{pmatrix} \tag{62}$$

We can eliminate P with the last equations and there remains :

$$(A + \frac{1}{\epsilon} BB^t)U = G \tag{63}$$

We solve this system by a standard method since the matrix $A + \epsilon^{-1} BB^t$ is symmetric positive definite, (but is still a big matrix). This method is easy to program because it can be shown that (63) is equivalent to

$$\alpha(u_h, v_h) + \nu(\nabla u_h, \nabla v_h) + \frac{1}{\epsilon} \sum_{1..M} (\nabla.u_h, q^i)(\nabla.v_h, q^i) = (f, v_h) \tag{64}$$

$$\forall v_h \in V_{oh} \quad u_h - u'_{\Gamma h} \in V_{oh}$$

It is useful to replace ϵ by ϵ/β_i, β_i being the area of the support of q^i; this corresponds to a penalization with the coefficients β_i in the diagonal of I in (62). Other penalizations have been proposed. A clever one (Hughes et al [118]) is to penalize by the equation for u with $v = \nabla q$:

$$(\nabla.u_h, q_h) = 0$$

is replaced by

$$(\nabla.u_h, q_h) + \epsilon[(\nabla p_h, \nabla q_h) - (\Delta u_h, \nabla q_h) - (f, \nabla q_h)] = 0$$

but the term $(\Delta u_h, \nabla q_h)$ can be dropped because $\nabla.u_h = 0$ (see (129) below).

5. ERROR ESTIMATION

(The presentation of this part follows closely Girault-Raviart [93] to which the reader is referred for a more detailed description of the convergence of finite element schemes for the Stokes problem.)

5.1. The abstract set up

Let us consider the following abstract problem

Find $u_h \in J_h$ where

$$J_h = \{u_h \in V_h : b(u_h, q_h) = 0 \quad \forall q_h \in Q_h\} \tag{65}$$

such that

$$a(u_h, v_h) = (f, v_h) \quad \forall v_h \in J_h \tag{66}$$

with the following notations :
- V_h sub-space of V, V Hilbert space, dim $V_h < +\infty$
- Q_h sub-space of Q, Q Hilbert space, dim $Q_h < +\infty$
- a(,) continuous bilinear of $V \times V \to$ R with

$$a(v, v) \geq \lambda ||v||_V^2 \quad \forall v \in V \text{ (V-ellipticity)} \tag{67}$$

- b(,) continuous bilinear of V × Q → R such that

$$0 < \beta \leq \inf_{q \in Q} \sup_{v \in V} \frac{b(v, q)}{||v||_V ||q||_Q} \tag{68}$$

- (,) duality product between V and (its dual) V'.
   We remark that these conditions are satisfied by the Stokes problem with

$$a(u, v) = \alpha(u, v) + \nu(\nabla u, \nabla v)$$

$$b(v, q) = -(\nabla.v, q)$$

$$V = H_0^1(\Omega)^n \quad Q = L^2(\Omega)/R$$

$$\lambda = inf\{\alpha, \nu\}.$$

$$(,) = \text{ scalar product of } L^2(\Omega).$$

$\beta = 1/C$ where C is such that for all q there is a u such that $\nabla.u = q$ and $|\nabla u|_0 \leq C|q|_0$

Remark:
   The notation in this formulation is somewhat misleading ; in fact J_{oh} and V_{oh} are now denoted J_h and V_h!

   The following proposition compares (66) with the continuous problem :

$$a(u, v) = (f, v) \quad \forall v \in J \tag{69}$$

$$u \in J = \{v \in V : \quad b(v, q) = 0 \quad \forall q \in Q\} \tag{70}$$

5.2. General theorems

Proposition 3
   If u_h and u are solutions of (65)-(66) and (69)-(70) respectively, and if V_h is non empty, then

$$||u - u_h||_V \leq C(\lambda)[\inf_{v_h \in J_h} ||u - v_h||_V + \inf_{q_h \in Q_h} ||p - q_h||_Q] \qquad (71)$$

Proof

We can show (Brezzi [39]) thanks to (68) that (69)-(70) is equivalent to the following problem : find $\{u, p\} \in V \times Q$ such that

$$a(u, v) + b(v, p) = (f, v) \quad \forall v \in V \qquad (72)$$

$$b(u, q) = 0 \quad \forall q \in Q \qquad (73)$$

By replacing v by $v_h \in J_h$ in (72) and by subtracting (66) from (72) we get

$$a(u - u_h, v_h) = -b(v_h, p - p_h) \quad \forall v_h \in J_h \qquad (74)$$

Or, by taking into account (65)

$$a(u - u_h, v_h) = -b(v_h, p - q_h) \quad \forall v_h \in J_h \quad \forall q_h \in Q_h \qquad (75)$$

and for all $w_h \in J_h$

$$a(w_h - u_h, v_h) = a(w_h - u, v_h) - b(v_h, p - q_h) \qquad (76)$$

So (cf (67) and the continuity of a and b)

$$\lambda ||w_h - u_h||_V \leq ||a|| ||w_h - u||_V + ||b|| ||p - q_h||_Q \qquad (77)$$

which implies

$$||u - u_h||_V \leq ||u - w_h||_V + ||w_h - u_h||_V \leq$$

$$(1 + ||a|| \lambda^{-1}) ||w_h - u||_V + \lambda^{-1} ||b|| ||p - q_h||_Q \quad \forall w_h \in J_h \quad \forall q_h \in Q \qquad (78)$$

Proposition 4 (Brezzi [39])

$$if \quad \inf_{q_h \in Q_h} \sup_{v_h \in V_h} \frac{b(v_h, q_h)}{||v_h||_V ||q_h||_Q} \geq \beta' > 0 \qquad (79)$$

$$then \quad \inf_{v_h \in J_h} ||u - v_h||_V \leq (1 + \frac{||b||}{\beta'}) \inf_{v_h \in V_h} ||u - v_h||_V \qquad (80)$$

Proof

Let us show first that if z_h is in the orthogonal space of J_h in V_h (denoted $J_h^\perp$) we have

$$\beta' \leq \frac{b(z_h, q_h)}{||z_h|| ||q_h||} \quad \forall q_h \in Q_h \quad \forall z_h \in J_h^\perp. \qquad (81)$$

This will be a direct consequence of (79) if we show the following :

$$z_h \in J_h^\perp \Rightarrow \exists q_h \text{ such that } z_h \text{ solution of } \sup_{z_h \in V_h} \left\{ \frac{b(z_h, q_h)}{||z_h||||q_h||} \right\}$$

In fact, the optimality conditions of the problem"Sup"are :

$$b(y_h, q_h) + \lambda(z_h, y_h) = 0 \quad \forall y_h \in V_h$$

with $\lambda = -||z_h||^{-3}$; since the conditions are necessary and sufficient for z_h to be a solution, it is sufficient to define q_h as a solution of

$$b(y_h, q_h) = (z_h, y_h) \quad \forall y_h \in V_h$$

which is possible because if B the operator defined by the above equation is not of maximal rank then by taking q_h in the kernel, we see that $\beta' = 0$ in (79).

Now let $v_h \in V_h$ be arbitrary. We can write $v_h = z_h + w_h$ where $w_h \in J_h$ and $z_h \in J_h^\perp$. Let us show that $||z_h|| \le C||u - v_h||$ for all u in J :

Since $z_h \in J_h^\perp$ we have

$$b(z_h, q_h) = b(v_h, q_h) = -b(u - v_h, q_h) \quad \forall q_h \in Q_h$$

Then we have from(81)

$$||z_h||_V \le \frac{1}{\beta'} \frac{b(z_h, q_h)}{||q_h||_Q} \le ||b||\beta'^{-1}||u - v_h||_V \tag{82}$$

so $\forall \ w_h \in J_h$, $\exists \ v_h \in V_h$ such that

$$||u - w_h||_V \le ||u - v_h||_V + ||z_h||_V \le (1 + \frac{||b||}{\beta'})||u - v_h||_V \tag{83}$$

Theorem 5
Let u_h and u be the solutions of (65)-(66) and (69)-(70)
If we have (67), (68) and (79) then

$$||u - u_h||_V \le C[\inf_{v_h \in V_h} ||u - v_h||_V + \inf_{q_h \in Q_h} ||p - q_h||_Q] \tag{84}$$

Proof
The proof follows directly from the 2 previous propositions.

Theorem 6
If p and p_h are the pressures associated with u and u_h (Lagrange multiplier of the constraints),we also have

$$||p - p_h||_Q \le C(\lambda)[\inf_{v_h \in V_h} ||u - v_h||_V + \inf_{q_h \in Q_h} ||p - q_h||_Q] \tag{85}$$

Proof :
From (74) and its corresponding discrete equation, we deduce that

$$b(v_h, p_h - q_h) = a(u - u_h, v_h) + b(v_h, p - q_h) \quad \forall v_h \in V_h \quad \forall q_h \in Q_h \quad (86)$$

So from (79)
$$\|p_h - q_h\|_Q \leq \beta'^{-1} \sup_{v_h \in V_h} \{ [a(u - u_h, v_h) + b(v_h, p - q_h)]/\|v_h\|_V \}$$

$$\leq \beta'^{-1}(\|a\|\|u - u_h\|_V + \|b\|\|p - q_h\|_Q).$$

We obtain :

$$\|p - p_h\|_Q \leq \beta'^{-1}\|a\|\|u - u_h\|_V + (\|b\|\beta'^{-1} + 1)\|p - q_h\|_Q \quad (87)$$

To study the generalized problem (64), we consider the following abstract problem where $c(p, q) = (Cp, q)$ is a symmetric $Q-$ elliptic continuous bilinear form :

Find $u_h^\epsilon \in V_h$ such that

$$a(u_h^\epsilon, v_h) + \epsilon^{-1}(C_h^{-1}B_h^T u_h^\epsilon, B_h^T v_h) = (f, v_h) \quad (88)$$

where C_h and B_h are defined by the relations

$$(C_h p_h, q_h) = c(p_h, q_h) \quad \forall p_h, q_h \in Q_h \quad (89)$$

$$(B_h^T u_h, q_h) = b(u_h, q_h) \quad \forall u_h \in V_h, \quad \forall q_h \in Q_h \quad (90)$$

Theorem 7
Under the hypothesis of theorem 5 and if β is independent of h, we have

$$\|u_h^\epsilon - u_h\|_V + \|p_h^\epsilon - p_h\|_Q \leq \epsilon C\|f\|_0 \quad (91)$$

Proof
We create an asymptotic expansion of u_h^ϵ and of p_h^ϵ

$$u_h^\epsilon = u_h^0 + \epsilon u_h^1 + \dots \quad (92)$$

$$p_h^\epsilon = p_h^0 + \epsilon p_h^1 + \dots$$

So (88) implies that

$$(C_h^{-1}B_h^T u_h^0, B_h^T v_h) = 0 \quad (93)$$

$$a(u_h^0, v_h) + (C_h^{-1}B_h^T u_h^1, B_h^T v_h) = (f, v_h), \quad (94)$$

$$a(u_h^1, v_h) + (C_h^{-1}B_h^T u_h^2, B_h^T v_h) = 0 \dots \forall v_h \in V_h \quad (95)$$

By putting $q_h = C_h^{-1} B_h^T v_h$, $p_h^0 = C_h^{-1} B_h^T u_h^1$, we see that u_h^0 is solution of the nonpenalized problem (66). We easily show that the u_h^i are bounded independently of h, and hence the theorem holds.

Remark:

If the inf sup condition is not verified (88) is still sensibles but the error due to penalization is $0(\sqrt{\epsilon})$ (cf. Bercovier [25]).

5.3. Verification of the inf-sup condition (79).

Finally for all the examples in section 3 if remains only to show the sufficient condition (79); we follow Nicolaides et al [181].

We shall do this for the element $P^1 bubble/P^1$. We recall the notation :

V represents $H_0^1(\Omega)^n$, $Q = L^2(\Omega)/R$, V_h and Q_h are approximations of the spaces ($P^1 bubble$, zero on the boundary, for V_h and P^1 for Q_h, both are assumed conforming). The lemma 1 below does not depend on the particular form of V_h and Q_h :

Lemma 1

The inf-sup condition (79) is equivalent to the existence of a Π_h

$$\Pi_h : V \to V_h \tag{96}$$

such that (C is independent of h):

$$(v - \Pi_h v, \nabla q_h) = 0 \quad \forall q_h \in Q_h, \quad \forall v \in V \tag{97}$$

$$||\Pi_h v||_1 \leq C||v||_1 \tag{98}$$

Proof

If Π_h exists, we have

$$\sup_{v_h \in V_h} \frac{(v_h, \nabla q_h)}{||v_h||_1} \geq \sup_{v \in V} \frac{(\Pi_h v, \nabla q_h)}{||\Pi_h v||_1} = \sup_{v \in V} \frac{(v, \nabla q_h)}{||\Pi_h v||_1} \tag{99}$$

$$\geq C^{-1} \sup_{v \in V} \frac{(v, \nabla q_h)}{||v||_1} \geq C^{-1} \beta |q_h|_o \quad (\beta' = C^{-1}\beta) \tag{100}$$

Conversely, let Π_h be an element in the orthogonal of J_h which satisfies

$$(\Pi_h v, \nabla q_h) = (v, \nabla q_h) \quad \forall q_h \in Q_h \tag{101}$$

then (cf. (81))

$$||\Pi_h v||_1 \leq \beta'^{-1} \frac{|(v, \nabla q_h)|}{|q_h|_0} \leq \beta'^{-1} ||v||_1 \tag{102}$$

because

$$|(v, \nabla q)| = |(\nabla . v, q)| \leq |\nabla . v|_0 |q|_0 \leq C||v||_1 |q|_0. \qquad (103)$$

Theorem 8

If the triangulation is regular (no angle tends to 0 or π when h tends to 0), then the $P^1 - bubble/P^1$ element satisfies the inf-sup condition with β' independent of h. So we have the following error estimate :

$$||u - u_h||_1 + |p - p_h|_o \leq Ch||u||_2. \qquad (104)$$

Proof
To simplify we assume $n = 2$. and we shall apply lemma 1. Let v be arbitrary in $H_0^1(\Omega)^n$.

Let $\Pi_h v$ be the interpolate defined by its values of the vertices q^i of the triangulation.

$$\Pi_h v(q^i) = v(q^i) \qquad (106)$$

- Here there is a technical difficulty because v may not be continuous and one must give a meaning to $v(q^i)$. We refer the reader to [12, lemma II.4.1] .

Nevertheless the values at the centre of the elements are chosen such that

$$\int_{T_k} \Pi_h v dx = \int_{T_k} v dx \quad \forall T_k \qquad (107)$$

Then we have

$$(\Pi_h v - v, \nabla q_h) = \sum_k [\int_{T_k} (\Pi_h v - v) dx] \nabla q_h|_{T_k} = 0 \qquad (108)$$

and we easily show that

$$||\Pi_h v||_1 \leq C||v||_1. \qquad (109)$$

Remark

1. This proof can be applied immediately to the case where the bubble is replaced by the element cut into three by the centre of a triangle.

2. It is not always easy to find a Π_h with (96)(98); another method is to prove (79) directly on a small domain made of a few elements and then show (Boland-Nicolaides [33]) that it implies (79) on the whole domain.

6. OTHER BOUNDARY CONDITIONS

6.1 Boundary Conditions without friction

The friction on the surface is given by $\nu(\nabla u + \nabla u^T)n - pn$. A no-friction condition on Γ_1 is written as

$$\nu(\nabla u + \nabla u^T)n - pn = 0 \text{ on } \Gamma_1 \subset \partial\Omega; \tag{110}$$

This condition may be useful if Γ_1 is an artificial boundary (such as the exit of a canal) or a free surface.

Let us consider the Stokes problem with (110) and

$$u = 0 \text{ on } \partial\Omega - \Gamma_1 \tag{111}$$

To include (110) in the variational formulation of the Stokes problem, we have to consider :

$$\alpha(u, v) + \frac{\nu}{2}(\nabla u + \nabla u^T, \nabla v + \nabla v^T) = (f, v) \quad \forall v \in J_{01}(\Omega) \tag{112}$$

$$u - u_\Gamma \in J_{01}(\Omega) = \{v \in J(\Omega) : v = 0 \text{ on } \partial\Omega - \Gamma_1\} \tag{113}$$

When $\nabla.u = 0$ we have :

$$(\nabla.\nabla u^T)_i = u_{j,ij} = 0 \tag{114}$$

and

$$(\nabla u + \nabla u^T, \nabla v^T) = (\nabla u + \nabla u^T, \nabla v) \tag{115}$$

So (112) implies

$$\alpha(u, v) - \nu(\Delta u, v) + \int_{\Gamma_1} \nu(\nabla u + \nabla u^T)nv = (f, v) \quad \forall v \in J_{01}(\Omega) \tag{116}$$

One can prove without difficulty that (116) implies (1) and (111).

The finite element approximation of (112)-(113) and the solution of the linear system is done as before. However in the saddle point algorithm the knowledge of the pressure does not decouple the velocity components (cf (51) in Algorithm 2).

6.2 Slipping Boundary Conditions.

It is also interesting to solve the Stokes problem without tangential friction and a slip condition on the velocity :

$$\tau.[\nu(\nabla u + \nabla u^T)n - pn] = 0, \quad u.n = 0 \text{ on } \Gamma_1 \subset \partial\Omega; \quad u = u_\Gamma \text{ on } \partial\Omega - \Gamma_1 \tag{117}$$

where τ is a vector tangential to $\partial\Omega$ (in 3D there are 2 conditions and 2 tangent vectors). We treat the problem in the same manner by solving .

$$\alpha(u, v) + \frac{\nu}{2}(\nabla u + \nabla u^T, \nabla v + \nabla v^T) = (f, v) \quad \forall v \in J_{0n1}(\Omega) \tag{118}$$

$$u - u_\Gamma \in J_{0n1}(\Omega) = \{v \in J(\Omega) : v = 0 \text{ on } \partial\Omega - \Gamma_1; \quad v.n|_{\Gamma_1} = 0\} \quad (119)$$

Here also the previous techniques can be applied ; however there are some additional problems in the approximation of the normal (cf. Verfürth [233]).

Some turbulence models replace the physical no slip condition on the surface of a solid by the slip condition :

$$au.\tau + \left(\frac{\partial u}{\partial n}\right).\tau = b, \quad u.n = 0 \text{ on } \Gamma_1 \subset \partial\Omega \quad (120)$$

where τ is a tangent vector to Γ_1 (two vectors in 3D).

We can either work with the standard variational formulation, that is :

$$\alpha(u, v) + \nu(\nabla u, \nabla v) + \int_{\Gamma_1} au.v = (f, v) + \int_{\Gamma_1} b\tau.v \quad \forall v \in J_{0n1}(\Omega) \quad (121)$$

$$u - u_\Gamma \in J_{0n1}(\Omega) = \{v \in J(\Omega) : v = 0 \text{ on } \partial\Omega - \Gamma_1; \quad v.n|_{\Gamma_1} = 0\} \quad (122)$$

or we can also reduce it to the previous case by noting that $n.\partial u/\partial \tau = -u.\tau/R$, so that (120) is equivalent to : ($a' = a+1/R$ where R is the radius of curvature):

$$a'u.\tau + \tau(\nabla u + \nabla u^T)n = b, \quad u.n = 0 \text{ on } \Gamma_1 \subset \partial\Omega$$

a condition which can be obtained with the variational formulation

$$\alpha(u, v) + \frac{\nu}{2}(\nabla u + \nabla u^T, \nabla v + \nabla v^T) + \int_{\Gamma_1} au.v = (f, v) + \int_{\Gamma_1} bv.\tau \quad (123)$$

$$\forall v \in J_{0n1}(\Omega) \quad u - u_\Gamma \in J_{0n1}(\Omega).$$

Finally, we note that

$$(\nabla q, v) = 0 \quad \forall q \in H^1(\Omega) \quad \Leftrightarrow \quad \nabla.v = 0 \text{ and } v.n|_\Gamma = 0$$

and so rather than working with J_{0n1} which is difficult to discretise, we can work with :

$$J'_{0n1}(\Omega) = \{v \in H^1(\Omega) : (\nabla q, v) = 0 \quad \forall q \in L^2(\Omega); \quad v|_{\partial\Omega-\Gamma_1} = 0\}$$

where the normals do not appear explicitly; but condition (120) is satisfied in the weak sense only. Parès [188] has shown that this is nevertheless a good method.

6.3. Boundary Conditions on the pressure and on the vorticity

To simplify the presentation, let us assume that Ω and $\partial\Omega$ are simply connected. Let us consider the Stokes problem with velocity equal to u_Γ on a part of the boundary Γ_2 and with prescribed pressure p^0 on the complementary part Γ_1 of $\partial\Omega$. If the tangential components of the velocity are prescribed on Γ_1 $(u \times n = u_\Gamma \times n)$ then the problem is well posed and we can solve it by considering the following variational formulation :

$$\alpha(u, v) + \nu(\nabla \times u, \nabla \times v) + \nu(\nabla.u, \nabla.v) = (f, v) + \int_{\Gamma_1} p^0 v.n \quad \forall v \in J_{0 \times n1}(\Omega)$$
(124)

$$u - u_\Gamma \in J_{0 \times n1}(\Omega) = \{v \in J(\Omega) : v = 0 \text{ on } \Gamma_2 = \partial\Omega - \Gamma_1; \quad v \times n|_{\Gamma_1} = 0\} \quad (125)$$

The finite element approximation of this space is carried out exactly as in chapter 2,§5

If we solve

$$\alpha(u, v) + \nu(\nabla \times u, \nabla \times v) + \nu(\nabla.u, \nabla.v) = (f, v) + \int_{\Gamma_1} d.v \quad \forall v \in J_{0n1}(\Omega) \quad (126)$$

$$u - u_\Gamma \in J_{0n1}(\Omega) = \{v \in J(\Omega) : v = 0 \text{ on } \Gamma_2 = \partial\Omega - \Gamma_1; \quad v.n|_{\Gamma_1} = 0\} \quad (127)$$

then we compute the solution of the Stokes problem with, on Γ_1 :

$$u.n = u_\Gamma.n, \quad n \times \nabla \times u = d, \quad \frac{\partial p}{\partial n} = f.n - \nabla.d, \quad\quad (128)$$

where u_Γ and d are arbitrary but such that $d.n = 0$.

Finally, V. Girault [91] has shown that the following problem

$$\alpha(u, v) + \nu(\nabla \times u, \nabla \times v) + \nu(\nabla.u, \nabla.v) = (f, v) \quad \forall v \in J'_{0n}(\Omega)$$

$$u - u_\Gamma \in J'_{0n}(\Omega) = \{v \in J(\Omega) : \quad v.n|_\Gamma = 0; n.\nabla \times v|_\Gamma = 0\},$$

yields

$$u.n = u_\Gamma.n, \quad n.\nabla \times u = 0, \quad \frac{\partial p}{\partial n} = f.n \text{ on } \Gamma$$

Naturally, we can 'mix' (125),(127) and the above condition on parts of Γ.

To sum up, we can treat a large number of boundary conditions but there are certain compatibilities needed. Thus it does not seem possible to have on the same boundary $u.n$ and p prescribed. Also it is not clear whether it is permissible to give all the components of $\nabla \times u$ on a part of the boundary in 3D. For more details, see Begue et al.[18].

6.4 Mixing different boundary conditions.

If we want to mix the boundary condition studied in §6.2 with those of §6.3 we are confronted with the choice of the variational formulation:

we have to use $(\nabla u, \nabla v)$ near to a boundary and $(\nabla \times u, \nabla \times v)$ on other parts of the boundary!

The following formulae are satisfied for all regular u, v and all Ω having a *polygonal* boundary :

$$(\nabla.u, \nabla.v) + (\nabla \times u, \nabla \times v) = (\nabla u, \nabla v) - \int_\Gamma [v\nabla u.n - v.n\nabla.u] \qquad (129)$$

$$\frac{1}{2}(\nabla u + \nabla u^T, \nabla v + \nabla v^T) = (\nabla.u, \nabla.v) + (\nabla u, \nabla v) + \int_\Gamma [v\nabla u.n - v.n\nabla.u] \quad (130)$$

We note that these formulae are still true if u, v are continuous and piecewise polygonal on a triangulation because the boundary integrals contain only the tangential derivatives of u and v or the jump of the tangential derivatives of piecewise continuous polynomial functions and those are zero at the discontinuous interfaces. Thus, (126) , for example, is *identical* to

$$\alpha(u, v) + \nu(\nabla u, \nabla v) - \frac{\nu}{2}\int_\Gamma [v\nabla u.n - v.n\nabla.u + u\nabla v.n - u.n\nabla.v] \qquad (131)$$

$$= (f, v) + \int_{\Gamma_1} d.v \quad \forall v \in J_{0n1}(\Omega) \quad u - u_\Gamma \in J_{0n1}(\Omega)$$

But the replacement of $(\nabla u + \nabla u^T, \nabla v + \nabla v^T)$ by $(\nabla u, \nabla v)$ produces an additional term $(\nabla.u, \nabla.v)$, which is zero in the continuous case and non zero in the discrete case.

There are further simplifications. Suppose, for example, that we have to solve the following problem in Ω with the simply connected polynomial boundary $\Gamma = \Gamma_1 \cup \Gamma_2 \cup \Gamma_3 \cup \Gamma_4$.

$$\alpha u - \nu\Delta u + \nabla p = f, \quad \nabla.u = 0 \text{ in } \Omega \qquad (132)$$

$$u|_{\Gamma_1} = u_\Gamma \qquad (133)$$

$$u.n = u_\Gamma.n; \quad au.\tau + \tau.\frac{\partial u}{\partial n} = b \text{ on } \Gamma_2 \qquad (134)$$

$$u \times n = u_\Gamma \times n; \quad p = p^0 \text{ on } \Gamma_3 \qquad (135)$$

$$u.n = u_\Gamma.n; \quad n \times \nabla \times u = d \text{ on } \Gamma_4 \qquad (136)$$

Then we could solve :

$$\alpha(u_h, v_h) + \nu(\nabla u_h, \nabla v_h) + a\int_{\Gamma_2} u_h.v_h \qquad (137)$$

$$= (f, v_h) + \int_{\Gamma_2} bv_h.\tau + \int_{\Gamma_3} p^0 v_h.n - \int_{\Gamma_4} d.v_h$$

$$+ \int_{\Gamma_3 \cup \Gamma_4} (v_h\nabla u_\Gamma.n - v_h.n\nabla.u_\Gamma) \quad \forall v_h \in J_h'$$

$$u_h - u_{\Gamma h} \in J'_h = \{v_h \in J_h : v_h|_{\Gamma_1} = 0, \quad v_h \times n|_{\Gamma_3} = 0, \quad v_h.n|_{\Gamma_2 \cup \Gamma_4} = 0\}$$

In fact, the extra integral is a function of u_Γ and not of u because it contains only tangential derivatives of $u.n$ on Γ_4 and on Γ_3 it contains only tangential derivatives of tangential components of u (in fact $\nabla.u - n\partial u/\partial n$).

Remark

Since these formulations apply only to polygonal domains one should be cautious of the "Babuska paradox" for curved boundaries;

the solutions obtained on polygonal domains may not converge to the right solution when the polygon converges to the curved boundary.

Figure 4.11 : *Iso values of pressure on a sphere in a Stokes flow.*
On the first figure (left) the computation was done with the P1/P1 invalid element (oscillations can be seen) on the second figure (right) the P1isoP2/P1 element was used (Courtesy of AMD-BA).

Chapter 5

Incompressible Navier-Stokes equations

1. INTRODUCTION

The Navier-Stokes equations :

$$u_{,t} + u\nabla u + \nabla p - \nu\Delta u = f \qquad (1)$$

$$\nabla .u = 0 \qquad (2)$$

govern Newtonian incompressible flows ; u and p are the velocity and pressure. These equations are to be integrated over a domain Ω occupied by the fluid, during an interval of time $]0,T[$. The data are :

- the external forces f,
- the viscosity ν (or the Reynolds number Re),
- the initial conditions at $t = 0$: u^0,
- the boundary conditions : u_Γ, for example.

These equations are particularly difficult to integrate for typical application (high Reynolds numbers) (small ν here) because of boundary layers and turbulence. Even the mathematical study of (1)-(2) is not complete ; the uniqueness of the solution is still an open problem in 3-dimensions.

There any many applications of the incompressible Navier-Stokes equations ; for example :

- heat transfer problems (reactors,boilers...)
- aerodynamics of vehicles (cars, trains, airplanes)
- aerodynamics inside motors (nozzles, combustion chamber...)
- meteorology, marine currents and hydrology.

Many tests problems have been devised to evaluate and compare numerical methods. Let us mention a few:

a) *The cavity problem*
The fluid is driven horizontally with velocity u by the upper surface of a cavity of size a (cf. figure 1). The Reynolds number is defined by : $R = ua/\nu$. (cf. Thomasset [229]). The domain of computation Ω is bidimensional.

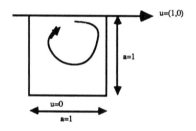

Figure 5.1

b) *The backward step :*
The parabolic velocity profile (Poiseuille flow) at the entrance and at the exit section are given, such that the flux at the entrance is equal to the flux at the exit. The domain is bidimensional and the Reynolds number is defined as $u_\infty l'/\nu$ where l' is the height of the step and u_∞ the velocity at the entrance at the centre of the parabolic profile (cf. Morgan et al [174]):

$$l = 3 \quad L = 22 \quad a = 1 \quad b = 1.5.$$

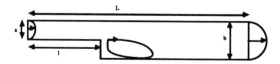

Figure 5.2 *The backward step problem (l'=1).*

c) *The cylinder problem.*

The problem is two dimensional until the Reynolds number rises above approximately 200 at which point it becomes tridimensional. The domain Ω is a periodic system of parallel cylinders of infinite length (for 3D computations one could take a cylinder length equal to 10 times their diameter). The integral, Q, of $u.n$ at the entrance of the domain is given and is equal to the total flux of the flow. The Reynolds number is defined by $3Q/2\nu$ (cf. Ronquist-Patera [202]).

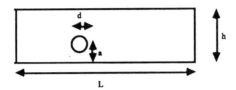

Figure 5.3

In this chapter, we give some finite element methods for the Navier-Stokes equations which are a synthesis of the methods in chapters 3 and 4. We begin by recalling the known main theoretical results. More mathematical details can be found in Lions [153], Ladyzhenskaya [138], Temam [228] ; details related to numerical questions can be found in Thomasset [229], Girault-Raviart [93], Temam [228] and Glowinski [95].

2. EXISTENCE, UNIQUENESS, REGULARITY.

2.1. The variational problem

We consider the equations (1)-(2), in $Q = \Omega \times]0, T[$ where Ω is a regular open bounded set in R^n, n = 2 or 3, with the following boundary conditions :

$$u(x,0) = u^0(x) \quad x \in \Omega \tag{3}$$

$$u(x,t) = u_\Gamma(x) \quad x \in \Gamma, \quad t \in]0, T[\tag{4}$$

We embed the variational form, as in the Stokes problem, in the space $J_o(\Omega)$:

$$J(\Omega) = \{v \in H^1(\Omega)^n : \nabla.v = 0\} : \tag{5}$$

$$J_o(\Omega) = \{u \in J(\Omega) : u|_\Gamma = 0\}. \tag{6}$$

The problem is to find $u \in L^2(O, T, J(\Omega)) \cap L^\infty(O, T, L^2(\Omega))$ such that

$$(u_{,t}, v) + \nu(\nabla u, \nabla v) + (u\nabla u, v) = (f, v), \quad \forall v \in J_o(\Omega) \tag{7}$$

$$u(0) = u^0 \tag{8}$$

$$u - u_\Gamma \in L^2(O, T, J_o(\Omega)) \tag{9}$$

In (7) (,) denotes the scalar product in L^2 in x as in chapter 4 ; the equality is in the L^2 sense in t. We take f in L^2 (Q) and u^0, u_Γ in $J(\Omega)$.

We search for u in L^∞ in order to ensure that the integrals containing $u\nabla u$ exist.

2.2. Existence of a solution.

Theorem 1
The problem (7)-(9) has at least one solution.

Proof (outline)
We follow Lions [153]; the outline of the proof only is given. To simplify, let us assume $u_\Gamma = 0$. Let $\{w^i\}_i$ be a dense basis V_s of $\{v \in D(\Omega)^n : \nabla.v = 0\}$ in H^s with $s = n/2$; this basis is defined by :

$$< w^i, v > = \lambda_i(w^i, v) \quad \forall v \in V_s. \tag{10}$$

-(the λ are arranged in increasing order, the notation $< >$ represents a scalar product in V_s and $D(\Omega)$ the set of C^∞ functions with compact support ; if $n = 2$, $\{w^i\}_i$ are the eigenvectors of the Stokes problem with $\alpha = \nu = 1$).

Let u^m be the approximated solution of (7)-(9) in the space generated by the first m solutions of (10) :

$$(u_t^m, w^j) + (u^m \nabla u^m, w^j) = (f, w^j) \quad j = 1..m \tag{11}$$

$$u^m = \sum_{i=1..m} v^i w^i \tag{12}$$

$$u^m(0) = u^{0m} \text{ the component of } u^o \text{ on } w^m . \tag{13}$$

On multiplying (11) by v^j and summing we get

$$\partial_t \frac{1}{2} |u^m|_o^2 + \nu |\nabla u^m|_o^2 = (f, u^m) \text{ because } (v\nabla v, v) = 0 \quad \forall v \in J_o(\Omega). \tag{14}$$

From (14) we deduce easily that

$$|u^m|_o^2 + 2\nu \int_o^t |\nabla u^m(\sigma)|^2 d\sigma \leq |u^{om}|_o^2 + C|f|_{0,Q} \tag{15}$$

which allows us to assert that the differential system in $\{v^i\}$ (11) has a solution on [O,T] and that solution is in a bounded set of $L^2(0, T, J_o(\Omega)) \cap L^\infty(O, T, L^2(\Omega))$.

Let P_m be the projection from $J_o(\Omega)'$ into the sub-space generated by $\{w^1, .. w^m\}$. From (1) we deduce that

$$u_t^m = P_m f - \nu P_m \Delta u^m - P_m(u^m \nabla u^m) \qquad (16)$$

P_m being a self-adjoint projector, its norm in time as an operator from V_s' in V_s', is less than or equal to 1. The mapping $\varphi \to \Delta\varphi$ being continuous from $J_0(\Omega)$ into $J_0(\Omega)'$, Δu is bounded in $L^2(O, T, J_0(\Omega)')$. Finally, by Sobolev embeddings, one shows that

$$\|u\nabla u\|_{(J_0)'} \leq C\|u\|_{L^p(\Omega)^n} \text{ with } p = \frac{2}{(1 - \frac{1}{n})} \qquad (17)$$

Which demonstrates that

$$u_t^m \text{ bounded in } L^2(O, T, J_0(\Omega)') \quad \forall m. \qquad (18)$$

By taking a subsequence such that u^m converges to u in $L^2(O, T, J_0(\Omega))$ weakly, $L^\infty(O, T, L^2(\Omega))$ weakly* , $L^2(Q)$ strongly and such that $u_t^m \to u_t$ in $L^2(O, T; J_0(\Omega)')$ weakly, one can pass to the limit in (11).

2.3. Uniqueness

Theorem 2
In two dimensions problem (7)-(9) has a unique solution.

Proof
The following relies on an inequality which holds if Ω is an open bounded set in R^2 : ($|\phi|_1$ denotes the semi norm in H_0^1 : $\phi \to (\nabla\phi, \nabla\phi)^{1/2}$).

$$\|\varphi\|_{L^4(\Omega)} \leq C|\varphi|_{0,\Omega}^{\frac{1}{2}} |\varphi|_{1,\Omega}^{\frac{1}{2}} \quad \forall v \in H_0^1(\Omega) \qquad (19)$$

because $(w\nabla u, w)$ is then bounded by $|w|_o\, |w|_1\, |u|_1$.
Let there be two solutions for (7)-(9) and let w be their difference.
We deduce from (7) with $v = w$:

$$\frac{1}{2}|w(t)|^2 + \nu \int_o^t |\nabla w|^2 dt = -\int_o^t (w\nabla u)w\, dt \qquad (20)$$

$$\leq C \int_o^t |w|_o|w|_1|u|_1 dt \leq \nu \int_o^t |w|_1^2 dt + C \int_o^t |w|_o^2|u|_1^2 d\sigma$$

So w is zero.

Theorem 3
If $|u^0|_1$ is small or if u is smooth (u $\in L^\infty(O, T ; L^4(\Omega))$ *then the solution of (7)-(9) is unique in three dimensions.*

Proof: See Lions [153].

Remark

It should be emphasized that theorem 3 says that if u is smooth, it is unique ; but one cannot in general prove this type of regularity for the solution constructed in theorem 1 ; so its uniqueness is an open problem. There are important reports which deal with the study of the possible singularities of the solutions. For instance, it is known that the singularities are "local" (Cafarelli et al. [46], Constantin et al. [59]) and we know their Hausdorff dimension is less than 1.

2.4. Regularity of the solution.

There are two important reasons for studying the regularity of the solutions of (7)-(9) :

1. As we have seen, the study of uniqueness is based on this regularity
2. The error estimates between the calculated solution and exact solution depend on the H^p norm of the latter. Unfortunately it is not unusual to have $u \in H^2$ but it is rather difficult to find initial conditions which give $u \in H^3$ (Heywood-Ranacher [111]) so it would appear useless from the point of view of classical error analysis, to approximate the Navier-Stokes equations with finite elements of degree 2 because they will not be more accurate than finite elements of degree 1; however even if there is some truth to what was said, this conclusion is too hasty and in fact one can show, on linear problems, that the error decreases with higher order elements even when there are singularities (but it does not decrease as much as it would if there were no singularities)..

In the case of two dimensions, there exists general theorems on regularity, for example :

Theorem 4 (Lions [153])
If f and $f_{,t}$ are in $L^2(Q)^2$, $f(.,0)$ in $L^2(\Omega)^2$, u^o in $J_0(\Omega) \cap H^2(\Omega)^2$ then the solution u of (7)-(9) is in $L^2(0,T;H^2(\Omega)^2)$.

As in 3 dimensions, as one cannot show uniqueness, the theorems are more delicate. Let us give an example of a result which, together with theorem 3 illustrates point 1.

The Hausdorff dimension d of a set O is defined as a limit (if it exists) of $\log N(\epsilon)/log(1/\epsilon)$ when $\epsilon \to 0$, where $N(\epsilon)$ is the minimum number of cubes having the length of their sides less than or equal to ϵ and which cover O. ($d = 0$ if O is a point, $d = 1$ if O is a curve of finite length, $d = 2$ for a surface... non integer numbers can be found in fractals (Feder [76])

Theorem 5 .(Cafarelli et al.[46])
Suppose that f is in $L^q(Q)^3$, $q > 5/2$, that $\nabla.f = 0$ and that $u^o.n$ and u_Γ are zero. Let u be a solution of (7)-(9) and S be the set of $\{x,t\}$ such that

$u \notin L^{\infty}_{loc}(D)$ *for all neighborhoods D of $\{x,t\}$. Then the Hausdorff dimension in space-time of S is less than 1.*

2.5. Behavior at infinity.

In general we are interested in the solution of (1)-(4) for large time because in practice the flow does not seem highly dependent upon initial conditions : the flow around a car for instance does not really depend on its acceleration history.

There are many ways "to forget" these initial conditions for a flow ; here are two examples :

    1. the flow converges to a steady state independent of t ;
    2. the flow becomes periodic in time.

Moreover, the possibility is not excluded that a little "memory" of initial conditions still remains ; thus in 1 the attained stationary state could change according to initial conditions since the stationary Navier-Stokes equations has many solutions when the Reynolds number is large (there is no theorem on uniqueness for ν large [28]). We distinguish other limiting states:

    3. quasi-periodic flow : the Fourier transform $t \rightarrow |u(x,t)|$ (where x is an arbitrary point of the domain) has a discrete spectrum.
    4.   Chaotic flows with strange attractors: $t \rightarrow |u(x,t)|$ has a continuous spectrum and the Poincaré sections (for example the points $\{u_1(x,nk), u_2(x,nk)\}_n$ for a given x) have dense regions of points filling a complete zone of space (in case 1, the Poincaré sections are reduced to a point when n is large, in cases of 2 and 3, the points are on a curve).

In fact, experience shows that the flows pass through the 4 regimes in the order 1 to 4 when the Reynolds number Re increases and the change of regime takes place at the bifurcation points of the mapping $\nu \rightarrow u$, where u is the stationary solution of (1)-(4). As in practice Re is very large (for us ν is small), 4 dominates. The following points are under study :

    a) Whether there exists attractors, and if so, can we characterize any of their properties ? (Hausdorff dimension, inertial manifold,...cf. Ghidaglia [89], Foias et al.[79] or Bergé et al [29] and the bibliography therein).
    b) Does $u(x,t)$ behave in a stochastic way and if so, by which law ? Can we deduce some equations for average quantities such as u, $|u|^2$, $|\nabla \times u|^2$... this is the problem of turbulence modeling and we shall cover it in a little more detail at the end of the chapter . (cf Lesieur [150], for example and the bibliography for more details).

Here are the main results relating to point a).

Consider (7)-(9)with $u_\Gamma = 0$, f independent of t, and Ω a subset of R^2. This system has an attractor whose Hausdorff dimension is between $cRe^{4/3}$ and CRe^2 where $Re = \sqrt{f}\ diam(\Omega)/\nu$ (cf. Constantin and al.[59], Ruelle [206][207]). These results are interesting because they give an upper bound for the number of points needed to calculate such flows (this number is proportional to $\nu^{-9/4}$).

In three dimensions, we do not know how to prove that (7)-(9) with the same boundary conditions has an attractor but we know that if an attractor exists and is (roughly) bounded by M in $W^{1,\infty}$ then its dimension is less than $CM^{3/4}\nu^{-9/4}$ (cf.[10]).

2.6. Euler Equations ($\nu = 0$)

As ν is small in practical applications, it is important to study the limit $\nu \to 0$. If $\nu = 0$ the equations (7)-(9) become Euler's equations. For the problem to be meaningful we have to change the boundary conditions ; so we consider the following problem :

$$u_{,t} + u\nabla u + \nabla p = f, \quad \nabla.u = 0 \tag{21}$$

$$u(0) = u^0 \tag{22}$$

$$u.n|_\Gamma = g \tag{23}$$

$$u(x) = u_\Gamma(x), \quad \forall x \in \Sigma = \{x \in \Gamma : \quad u(x).n(x) < 0\}. \tag{24}$$

Proposition 1

If Ω is bidimensional and simply connected, if $f \in L^2(Q)$, $u^0 \in H^1(\Omega)^2$, $g = u_\Gamma.n$ and $(u_\Gamma)_i$ are in $H^{1/2}$ (Γ) and the integral of g on Γ is zero, then (21) - (24) has a unique solution in $L^2(O,T ; H^1(\Omega)^2) \cap L^\infty(O,T,L^2(\Omega)^2)$.

Proof (sketch)
To simplify, we assume $f = 0, g = 0$. Since $\nabla.u = 0$ there exists a ψ such that :

$$u_1 = \psi_{,2}, u_2 = -\psi_{,1}. \tag{25}$$

By taking the curl of (21)-(22) and by putting

$$\omega = u_{1,2} - u_{2,1} = -\Delta\psi \tag{26}$$

We see that

$$\omega_t + u\nabla\omega = 0 \tag{27}$$

$$\omega(0) = u^0_{1,2} - u^0_{2,1} = \nabla \times u^0 \tag{28}$$

$$\omega(x) = \omega_\Gamma(x) \text{ (derived from g and } u_\Sigma \text{ ; here } \omega_\Gamma \text{ is zero) if } x \in \Sigma. \tag{29}$$

If u is continuous, Lipschitz and g is zero, (27)-(28) can be integrated by the method of characteristics (Cf. Chapter 3) and ω has the same smoothness as that of $\nabla \times u^0$. So we have an estimate which relates $||\omega||_s$ to the Holderian norm C^0 of u (Kato [129])

$$||\omega|| \leq C(||u|| + ||\nabla \times u^0||) \tag{30}$$

Moreover, from (25)-(26), (23) with $g = 0$ gives

$$\psi|_\Gamma = 0, \tag{31}$$

Thus from (26), we have

$$||u||_{1+s} \leq C||\omega||_s. \tag{32}$$

So it is sufficient to construct a sequence in the following way :
$$u^n \text{ given} \Rightarrow \omega^{n+1} \text{ calculated by (27)-(29)}$$
$$\omega^{n+1} \text{ given} \Rightarrow u^{n+1} \text{ calculated by (25),(26),(30)}$$
and to use the estimates (30) and (32) to pass to the limit.
A complete proof can be seen in Kato [129], McGrath[170], Bardos[16].

Remark 1
In three dimensions we only have an existence theorem in the interval $]O, \tau[$, τ small. In fact, equation (27) becomes

$$\omega_{,t} + u\nabla\omega - \omega\nabla u = 0$$

which can give terms in $e^{\lambda t}$ where λ are the eigenvalues of ∇u.
A number of numerical experiments have shown the formation of singularities after a finite time but the results have not yet been confirmed theoretically (Frisch [84], Chorin [52], Sulem [225]).

Remark 2
We know little about convergence of the solution of the Navier-Stokes equation towards the solution of Euler's equation when $\nu \rightarrow 0$. In stationary "laminar" cases, one can prove existence of boundary layers of the form $e^{-y/\sqrt{\nu}}$ (y being the distance to the boundary) in the neighborhood of the walls (Landau-Lifschitz [141], Rosenhead [203]) ; so convergence is in L^2 at most in those cases.

3. SPATIAL DISCRETISATION

3.1. Generalities.

The idea is simple: we discretise in space by replacing $J_o(\Omega)$ by J_{oh} in (7)-(9):

Find $u_h \in J_h$ such that

$$(u_{h,t}, v_h) + (u_h \nabla u_h, v_h) + \nu(\nabla u_h, \nabla v_h) = (f, v_h) \quad \forall v_h \in J_{oh} \qquad (33)$$

$$u_h(0) = u_h^0 \quad u_h - u_{\Gamma_h} \in J_{oh} \qquad (34)$$

If J_{oh} is of dimension N and if we construct a basis for J_{oh} then (33)-(34) becomes a nonlinear differential system of N equations:

$$AU' + B(U,U) + \nu CU = G. \qquad (35)$$

One could use existing library programs (LINPAK[152] for example) to solve (35) but they are not efficient, in general, because the special structure of the matrices is not used. This method is known as the "Method of Lines"

So we shall give two appropriate methods, taking into account the following two remarks :

$1°$) If $\nu >> 1$ (33)-(34) tends towards the Stokes problem studied in Chapter 4.

$2°$) If $\nu = 0$ (33)-(34) is a non-linear convection problem in J_{oh}. In particular in 2-D the convection equation (27) is underlying the system so the techniques of chapter 3 are relevant.

As we need a method which could adapt to all the values of ν we shall *to take the finite elements in the family studied in the Chapter 4 and the method for time discretisation adapted to convection studied in Chapter 3.* But before that, let us see a convergence theorem for the approximation (33), (35).

3.2. Error Estimate .

Let us take the framework of Chapter 4: $J(\Omega)$ is approximated by

$$J_h = \{v_h \in V_h : \quad (\nabla . v_h, q_h) = 0 \quad \forall q_h \in Q_h\} \qquad (36)$$

and $J_o(\Omega)$ is approximated by $J_{oh} = J_h \cap H_0^1(\Omega)^n$. We put :

$$V_{oh} = \{v_h \in V_h : \quad v_h|_{\Gamma_h} = 0\}. \qquad (37)$$

Assume that $\{V_{oh}, Q_h\}$ is such that
- there exists $\Pi_h : V_o^2 = H^2(\Omega)^n \cap H_0^1(\Omega)^n \rightarrow V_{oh}$ such that

$$(q_h, \nabla.(v - \Pi_h v)) = 0 \quad \forall q_h \in Q_h \quad ||v - \Pi_h v|| \leq Ch||v||_{2,\Omega}. \qquad (38)$$

- there exists $\pi_h : H^1(\Omega) \rightarrow Q_h$ such that

$$|q - \pi_h q|_{0,\Omega} \leq Ch||q||_{1,\Omega} \qquad (39)$$

$$\inf_{q_h \in Q_h} \sup_{v_h \in V_{oh}} \frac{(\nabla.v_h, q_h)}{|q_{h\pi}|_0 ||v_h||_1} \geq \beta$$

$$||v_h||_1 \leq \frac{C}{h} |v_h|_0 \quad \forall v_h \in V_{0h} \text{ (inverse inequality)}$$

Remark :
These conditions are satisfied by the $P^1 + bubble/P^1 element$ and among others, the element $P^1 iso\ P^2/P^1$.

To simplify, let us assume that

$$u^o = 0 \quad and \quad u_\Gamma = 0. \tag{40}$$

Theorem 5. (Bernardi-Raugel [31])
a) $n=2$. If the solution of (7)-(9) is in $L^2(0,T; H_0^1(\Omega)^2) \cap C^0(0,T; L^q(\Omega)^2)$, $q > 2$, then problem (33)-(34) has only one solution for h small satisfying

$$||u - u_h||_{L^2(o,T;H_0^1(\Omega))} \leq C(\nu)h||u\nabla u||_{0,Q} \tag{41}$$

b) $n=3$. If the norm of $\nu^{-1}u$ in $C^o(0,T; L^3(\Omega)^3)$ is small and if $u \in L^2(0,T; H^2(\Omega)^3) \cap H^1(0,T; L^2(\Omega)^3)$ then (34)-(36) has a unique solution with the same error estimate as in a).

Proof (sketch)
The proof is technical, but it is based on a very interesting idea (introduced in Brezzi et al [38]).
We define the operator

$$G : u \rightarrow G(u) = u\nabla u - f. \tag{42}$$

We observe that the non stationary Stokes problem

$$u_{,t} - \nu\Delta u + \nabla p = g, \quad \nabla.u = 0, \quad u(0) = 0, \quad u_\Gamma = 0 \tag{43}$$

defines a linear map $L : g \rightarrow u$ and that the solution of (7)-(9) is nothing but the zero of $u \rightarrow u + LG(u)$:

$$u \text{ solution of (7)-(9)} \quad \Leftrightarrow \quad F(u) = u + LG(u) = 0. \tag{44}$$

Similarly, for the discrete problem if L_h is the nonstationary discrete Stokes operator underlying (33)-(34):

$$u \text{ solution of (33)-(34)} \quad \Leftrightarrow \quad F_h(u) = u_h + L_h G(u_h) = 0. \tag{45}$$

To prove the existence of a solution for the discrete problem, we note that the derivative of F_h with respect to u, $F_h'(u_h)$, is a linear map (because F_h is quadratic) and that it is an isomorphism if v_h is near to u. Then we see that the solution is also a fixed point of

$$v \to \varphi_h(v) = v - F_h'(v_h)^{-1}F_h(v) \qquad (46)$$

for all v_h near to u. For $\varphi_h' = I - F'(v_h)^{-1}F'(v)$ in the neighborhood of 0 is a contraction and from a fixed point theorem we have the existence and uniqueness of the solution.

To obtain an estimate, we calculate the following identity :

$$||u - u_h|| = ||F_h'^{-1}[F_h'.(u - u_h) - (F_h(u) - F_h(u_h)) + F_h(u)]|| \qquad (47)$$

$$\leq ||F_h'^{-1}[\int_0^1 [F_h'(u_h) - F_h'(u_h + \theta(u - u_h))].(u - u_h)d\theta] + F_h(u)||$$

$$\leq ||F_h'^{-1}||[\sup_v ||F_h'(u_h) - F_h'(v)||||u - u_h|| + ||F_h(u)||]$$

So, for small h we have :

$$||u - u_h|| \leq C||F_h(u)|| \qquad (48)$$

Moreover

$$F_h(u) = u + L_h G(u) = u + LG(u) + (L_h - L)G(u) = (L_h - L)G(u) \qquad (49)$$

but $||(L_h - L)G(u)||$ is precisely the error in the resolution of the non-stationary Stokes problem with $g = G(u)$. Hence the result. For more details, see [4][5] [16,IV§3]

Remark 1
The proof of theorem 5 suggests that the usual inf sup condition on the conforming finite elements in the error estimate for the non-stationary Stokes problem is necessary for Navier-Stokes equations.

Remark 2 .
The error estimate of theorem 5 shows that h should be small when ν is small. We note that $||u\nabla u||$ depends also on ν. For example in a boundary layer in $e^{-y/\sqrt{\nu}}$,

$$|u\nabla u|_0 \approx (\int_0^1 \nu^{-1}e^{-\frac{4y}{\sqrt{\nu}}}dy)^{\frac{1}{2}} = O(\nu^{-\frac{1}{4}}) \qquad (50)$$

Since $C(\nu) \approx 0(\nu)$, we should take $h \ll \nu^{5/4}$. The same reasoning leads to $h \ll \nu^{5/8}$ if we want $|u - u_h|_{1,Q} \ll 1$. We note that the numbers are less pessimistic than those given by the attractor argument. This is not contradictory as there must exist attractor modes of modulus less than $\nu^{-5/8}$.

4. TIME DISCRETISATION

4.1 Semi-explicit discretisation in t

Let us take the simplest explicit scheme studied in Chapter 3, that is the Euler scheme.

Scheme 1:
Find u_h^{n+1} with $u_h^{n+1} - u_{\Gamma h} \in J_{0h}$ such that

$$\frac{1}{k}(u_h^{n+1}, v_h) = g(v_h) \quad \forall v_h \in J_{0h}, \tag{51}$$

$$g(v_h) = -\nu(\nabla u_h^n, \nabla v_h) + (f^n, v_h) + \frac{1}{k}(u_h^n, v_h) - (u_h^n \nabla u_h^n, v_h) \tag{52}$$

Evidently, k has to satisfy a stability condition of the type

$$k < C(u, h, \nu) \tag{53}$$

As for the Stokes problem (51) is equivalent to

$$\frac{1}{k}(u_h^{n+1}, v_h) + (\nabla p_h, v_h) = g(v_h) \quad \forall v_h \in V_{0h} \tag{54}$$

$$(\nabla . u_h^{n+1}, q_h) = 0 \quad \forall q_h \in Q_h; \tag{55}$$

it is still necessary to solve a linear system to get u_h^{n+1}; we use a mass-lumping formula and the $P^1 isoP^2/P^1 element$, with quadrature points at the nodes $\{q^i\}_i$ of the finer *triangulation*; σ_i denotes the area of support of the P^1 continuous basis functions, v^i, associated to node q^i; we replace (54) by

$$\frac{\sigma_i}{3k}u_i^{n+1} + (\nabla p_h, v^i) = g(v^i) + \frac{\sigma_i}{3k}u_i^n - \frac{1}{k}(u_h^n, v^i) \quad \forall i \tag{56}$$

$$(\nabla . u_h^{n+1}, q_h) = 0 \quad \forall q_h \in Q_h \quad u_h^{n+1}(x) = \sum_i u_i v^i(x). \tag{57}$$

We can now substitute u_h in (57) and we obtain

$$\sum_i \frac{3k}{\sigma_i}(\nabla p_h, v^i)(\nabla q_h, v^i) = G(q_h) \quad \forall i. \tag{58}$$

that is a linear system of a Laplacian type.

The main drawback of these types of methods is the stability (53).

There exist schemes which are almost unconditionally stable such as the rational Runge-Kutta used by Satofuka [212]:

To integrate

$$V_{,t} = F(V) \tag{59}$$

we use

Scheme 2

$$V^{n+1} = V^n + [2g^1(g^1, g^3) - g^3(g^1, g^1)](g^3, g^3)^{-1} \qquad (60)$$

where $(.,.)$ represents the scalar product in the space of $V(t)$ and

$$g^1 = kF(V^n)$$

$$g^2 = kF(V^n - cg^1)$$

$$g^3 = bg^1 + (1 - b)g^2$$

This method is of order 2 if $2bc = -1$ and of order 1 if it is not. For a linear equation with constant coefficients, the method is stable when $2bc \leq -1$ (for the proof, it is sufficient to calculate the amplification coefficient when it is applied to (59) and $F(V) = FV$ where F is a diagonal matrix). For our problem, we must add a step of spatial projection on zero divergence functions of J_{0h}. That is, we solve :

$$\frac{\sigma_i}{3k} u_i^{n+1} + (\nabla p_h, v^i) = \frac{\sigma_i}{3k} v_i^{n+1} \quad \forall v_h \in V_{0h} \qquad (61)$$

$$(\nabla . u_h^{n+1}, q_h) = 0 \quad \forall q_h \in Q_h \qquad (62)$$

Numerical experiments using this method for the Navier-Stokes equations discretised with the $P^1 isoP^2/P^1$ finite elements can be found in Singh [215].

4.2. Semi-Implicit and Implicit discretisations

A semi-implicit discretisation of O(k) for (34)-(36) is

Scheme 3:

$$\frac{1}{k}(u_h^{n+1}, v_h) + \nu(\nabla u_h^{n+1}, \nabla v_h) + (u_h^n \nabla u_h^{n+1}, v_h) = (f^{n+1}, v_h) + \frac{1}{k}(u_h^n, v_h) \quad (63)$$

A version $O(k^2)$ is easy to construct from the Crank-Nicolson scheme. This scheme is very popular because it is conceptually simple and almost fully implicit : yet it is not unconditionally stable and each iteration requires the solution of a non symmetric linear system. Thus on the cavity problem with 10×10 mesh and $k = 0.1$ the method works well until $\nu \approx 1/500$, beyond which oscillations develop in the flow.

Let us consider the following implicit Euler scheme of order O(k)

Scheme 4:

$$\frac{1}{k}(u_h^{n+1}, v_h) + \nu(\nabla u_h^{n+1}, \nabla v_h) + (u_h^{n+1} \nabla u_h^{n+1}, v_h) = (f^{n+1}, v_h) + \frac{1}{k}(u_h^n, v_h) \quad (64)$$

$\forall v_h \in J_{0h}$. This scheme is unconditionally stable but we must solve a non-linear system.

4.3 Solution of the non-linear system (59) .

This non-linear system can be solved by Newton's method, by a least square method in H^{-1} or by the GMRES algorithm :

4.3.1 Newton's Method

The main loop of the algorithm is as follows :
For $p = 1..pMax$ do:
1. Find δu_h with

$$(\delta u_h, v_h)\frac{1}{k} + \nu(\nabla \delta u_h, \nabla v_h) + (u_h^{n+1,p} \nabla \delta u_h + \delta u_h \nabla u_h^{n+1,p}, v_h) = \quad (65)$$

$$-\{(u_h^{n+1,p}, v_h)\frac{1}{k} + \nu(\nabla u_h^{n+1,p}, \nabla v_h) + (u_h^{n+1,p} \nabla u_h^{n+1,p}, v_h)$$

$$-(f^{n+1}, v_h) - (u_h^n, v_h)\frac{1}{k}\} \quad \forall v_h \in J_{oh} \quad \delta u_h \in J_{oh} \quad (66)$$

2. Put

$$u_h^{n+1,p+1} = u_h^{n+1,p} + \delta u_h \quad (67)$$

But here also, experience shows that a condition between k, h and ν is necessary for the stability of the scheme because if ν is very small with respect to h, problem (59) has many branch of solutions or because the convergence conditions for Newton's method (hessien > 0) are not verified. Again for a cavity with a 10×10 and $k = 0.1$ we can use this scheme till Re=1000 approximately. If the term $u\nabla u$ is *upwinded* , one can get a method which is unconditionally stable for all mesh k ; this is the object of a few methods studied below.

4.3.2 Abstract least-squares and θ scheme.

This method has been introduced in Chapter 2, Paragraph 4.2. It consists here of taking for u_h^{n+1} the solution of the problem

$$\min_{w_h - u_\Gamma \in J_{oh}} \{\int_\Omega |u_h - w_h|^2 dx : \quad u_h - u_\Gamma \in J_{oh}, \quad (68)$$

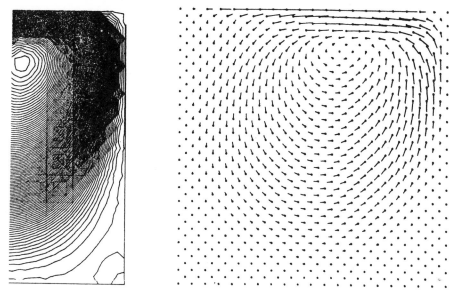

Figure 5.4 : *Streamline of a bidimensional flow in a cavity. The computation was done with the P1isoP2/P1 element at Reynolds 500 with a uniform triangulation of 400 triangles and abstract least squares without upwinding. (Courtesy of AMD-BA)* .

$$\frac{1}{k}(u_h, v_h) + \nu(\nabla u_h, \nabla v_h) + (w_h \nabla w_h, v_h) = (f^{n+1}, v_h) + \frac{1}{k}(u_h^n, v_h) \quad \forall v_h \in J_{0h}\}$$

The solution of this problem by a preconditioned conjugate gradient method, solving three Stokes problems per iteration (calculation of the step size in the descent directions is analytic because the optimized quantity is of degree 4). This method has good stability qualities. Only symmetric linear systems are solved but it is more costly than other methods ; in Bristeau et al [44] a set of numerical experiments can be found.

To decrease the cost Glowinski [96] suggested the use of a θ scheme because it is second order accurate for some values of the parameters and because it allows a decoupling between the convection and the diffusion operators.

If B is the convection operator $u_h \nabla$ and A is the Stokes operator the θ scheme is:

$$\frac{1}{k\theta}(u^{n+\theta} - u^n) + Au^{n+\theta} + Bu^n = f^{n+\theta}$$

$$\frac{1}{(1-2\theta)k}(u^{n+1-\theta} - u^{n+\theta}) + Au^{n+\theta} + Bu^{n+1-\theta} = f^{n+1-\theta}$$

$$\frac{1}{k\theta}(u^{n+1} - u^{n+1-\theta}) + Au^{n+1} + Bu^{n+1-\theta} = f^{n+1}$$

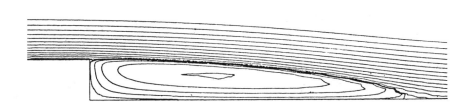

Figure 5.5 : *Streamlines for the backward step problem with the P2/P1 element, abstract least squares without upwinding at Reynolds 191. (Courtesy of AMD-BA)* .

So the least square method will be used only for the central step while the two other steps are generalized Stokes problems. Alternatively, the convection step can be solved by other methods such as GMRES.

4.3.3 The GMRES algorithm.

The GMRES algorithm (Generalized Minimal RESidual method) is a quasi-Newtonian method deviced by Saad [210]. The aim of quasi-Newtonian methods is to solve non-linear systems of equations :

$$F(u) = 0, \quad u \in R^N$$

The iterative procedure used is of the type :

$$u_{n+1} = u_n - (J_n)^{-1}F(u_n)$$

where J_n is an approximation of $F'(u_n)$. To avoid the calculation of F' the following approximation may be used:

$$D_\delta F(u; v)\frac{\equiv F(u + \delta v) - F(u)}{\delta} \cong F'(u)v. \tag{69}$$

As in the conjugate gradient method to find the solution v of $Jv = -F$ one considers the Krilov spaces :

$$K_n = Sp\{r^0, Jr^0..., J^{n-1}r^0\}$$

where $r^0 = -F - Jv^0$, v^0 is an approximation of the solution to find. In GMRES the approximated solution v^n used in (69) is the solution of

$$\min_{v \in K_n} ||r^0 - J(v - v^0)||.$$

The algorithm below generates a quasi-Newton sequence which $\{u_n\}$ hopefully (it is only certain in the linear case) will converge to the solution of $F(u) = 0$ in R^N :

Algorithm (GMRES) :

0 . *Initialisation:* Choose the dimension k of the Krylov space; choose u_0. Choose a tolerance ϵ and an increment δ , choose a preconditioning matrix $S \in R^{N \times N}$. Put $n = 0$.

1. a. Compute with (69) $r_n^1 = -S^{-1}(F_n + J_n u_n)$, $w_n^1 = r_n^1/ \, ||r_n^1||$, where $F_n = F(u_n)$ and $J_n v = D_\delta F(u_n; v)$.
    b. For $j = 2..., k$ compute r_n^j and w_n^j from

$$r_n^j = S^{-1}[D_\delta F(u_n; w_n^{j-1}) - \sum_{i=1}^{j-1} h_{i,j-1}^n w_n^i]$$

$$w_n^j = \frac{r_n^j}{||r_n^j||}.$$

where $h_{i,j}^n = w_n^{iT} S^{-1} \, D_\delta F(u_n; w_n^j)$
    2. Find u_{n+1} the solution of

$$\min_{v=u_n+\sum_0^k a_j w_j^n} \, ||S^{-1}F(v)||^2$$

$$\cong \min_{a_1,a_2,...a_k} \, ||S^{-1}[F(u_n) + \sum_{j=1}^k a_j D_\delta F(u_n; w_n^j)]||^2$$

3. If $||F(u_{n+1})|| < \epsilon$ stop else change n into n+1.

The implementation of Y. Saad also includes a back-tracking procedure for the case where u_n departs too far away from the solution. The result is a black box for the solution of system of equations which needs only a subprogram to calculate $F(u)$ given u; there is no need to calculate F'. Experiments have shown that this implementation is very efficient and robust.

5. DISCRETISATION OF THE TOTAL DERIVATIVE.

5.1. Generalities

Let us apply the techniques developed in chapter 3:

$$u_{,t} + u\nabla u \approx \frac{1}{k}(u^{n+1} - u^n o X^n)$$

we obtain the following scheme :

Scheme 5:

$$\frac{1}{k}(u_h^{n+1}, v_h) + \nu(\nabla u_h^{n+1}, \nabla v_h) = (f^n, v_h) + \frac{1}{k}(u_h^n o X_h^n, v_h) \quad \forall v_h \in J_{0h} \quad (70)$$

where $X_h^n(x)$ is an approximation of the foot of the characteristic at time nk which passes through x at time $(n+1)k$ $(X_h^n(x) \approx x - u_h^n k)$.

We note that the linear system in (70) is still symmetric and independent of n. If X_h^n is well chosen, this scheme is unconditionally stable, and convergent in $0(k + h)$.

Theorem 6 :
If

$$X_h^n(\Omega) \subset \Omega, \quad |det(\nabla X_h^n)^{-1}| \leq 1 + Ck, \qquad (71)$$

then scheme 5 is unconditionally stable.

Proof :
This depends upon the change of variable $y = X_h^n(x)$ in the integral

$$\int_\Omega \phi(X_h^n(x))dx = \int_{X_h^n(\Omega)} \phi(y)det(\nabla X_h^n)^{-1}dy \qquad (72)$$

From this we obtain

$$|u_h^n o X_h^n|_{0,\Omega} \leq |det(\nabla X_h^n)^{-1}|_{\infty,\Omega}|u_h^n|_{0,\Omega} \qquad (73)$$

In getting a bound from (70) with $v_h = u_h^{n+1}$ and by using (73) and hypothesis (71), we obtain

$$||u_h^{n+1}||_\nu^2 \equiv |u_h^{n+1}|_0^2 + k\nu|\nabla u_h^{n+1}|_0^2 \leq (k|f^n|_0 + (1 + Ck)|u_h^n|_0)|u_h^{n+1}|_0 \quad (74)$$

Since $|u_h|_0 \leq ||u_h||_\nu$, we obtain

$$||u_h^{n+1}||_\nu \leq k|f^n|_0 + (1 + Ck)||u_h^n||_\nu; \qquad (75)$$

or, after summation,

$$||u_h^{n+1}||_\nu \leq k\sum_{i\leq n}|f^i|_0(1 + Ck)^{n-i} + ||u_h^0||_\nu(1 + Ck)^n \qquad (76)$$

and hence the result follows with $n \leq T/k$.

Remark :
If $u_h^n = \nabla \times \psi_h$, with ψ_h P^1 piecewise and continuous and if X_h^n is the exact solution of

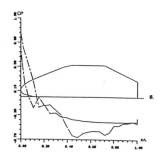 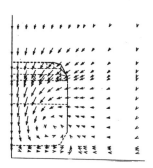

The pressure on the car at the mid-plane.   *The vortex behing the car.*

Figure 5.6 : *3D flow around a car with the P1bubble/P1 element at Reynolds 500. The figures show the trace of the mesh on the car, the velocities on a plane perpandicular to the flow behing half the car and the pressure on the car at the plane of symetry. (Computed by F. Hecht).*

$$\frac{d}{d\tau}X_h(\tau) = u_h^n(X_h(\tau)) \tag{77}$$

then (71) is satisfied with $C = 0$. It is theoretically possible to integrate (77) exactly since X_h is then a piecewise straight line but, in practice, the exact calculation of the integral $(u_h^n o X_h^n, v_h)$ is unnecessarily costly ; we use a direct Gauss formula (Pironneau [190])

$$(u_h^n o X_h, v_h) \cong \sum_i u_h(X_h^n(\xi^i))v_h(\xi^i)\pi_i \tag{78}$$

or a dual formula (Benqué et al. [22])

$$(u_h^n o X_h, v_h) \cong \sum_i u_h(\xi^i)v_h(X_h^{n-1}(\xi^i))\pi_i \tag{79}$$

5.2. Error analysis with P1 bubble/P1 .

Definition of the method :
To simplify, we assume that $u_\Gamma = 0$; we choose

$$J_{oh} = \{v_h \in V_{oh} : (\nabla.v_h, q_h) = 0, \forall q_h \in Q_h\} \tag{80}$$

$$V_{oh} = \{v_h \in C^o(\Omega) : v_{h|T} \in P^1, \forall T \in T_{\frac{h}{2}}\} \tag{81}$$

$$Q_h = \{q_h \in C^o(\Omega) : q_{h|T} \in P^1, \forall T \in T_h\}. \tag{82}$$

where T_h is a regular triangulation of Ω (assumed to be polygonal) and $T_{h/2}$ is the triangulation obtained by dividing each triangle (resp. tetrahedron) in 3 (resp. 4) by joining the center of gravity to the vertices (fig. 4.2).

We define $X_h^n(x)$ as the extremity of the broken line $[\chi^o = x, \chi^1, ..., \chi^{m+1} = X_h^n(x)]$ such that the χ^j are the intersections with the sides (faces), χ^{j-1} is calculated from χ^j in the direction $-u_h^m(\chi^j)$ and the time covered from x to χ^{m+1} is k :

$$\text{For each j } \exists \rho > 0 \text{ such that } \chi^{j+1} = \chi^j + \rho u_h^n(\chi^j)|_{T_j} \in \partial T_j, \qquad (83)$$

$$\sum_{j=0..m} \frac{|\chi^{j+1} - \chi^j|}{|u_h^n(\chi^j)|_{T_j}} = k \qquad (84)$$

In practice, we use (78) or (79) with the 3 (resp. 4) Gauss points inside the triangles, but with this integral approximation, the error estimation is an open problem; furthermore (71.b) is not easily satisfied, so we shall define another X_h^n.

Definition of X_h^n satisfying (71) :
Let ψ_h^n be the solution of

$$(\nabla \times \psi_h^n, \nabla \times w_h) + (\nabla.\psi_h^n, \nabla.w_h) = (u_h^n, \nabla \times w_h), \forall w_h \in W_h, \qquad (86)$$

$$\psi_h^n \in W_h = \{w_h \in V_h : w_h \times n|_\Gamma = 0, \int_{\Gamma_i} w_h.n d\gamma = 0\} \qquad (87)$$

where Γ_i are the connected components of Γ. Let us assume for simplicity that Ω is simply connected. This problem has already been studied in paragraph 5 of chapter 2 (here however $w_h \times n$ is zero on Γ instead of $w_h \times n_h$; W_h is non empty because Γ is a polygon . We have seen that (86)-(87) has one and only one solution and that $\nabla \times \psi_h^n \cong u_h^n$. Since $\nabla \times \psi_h^n$ is piecewise constant and with zero divergence in the sense of distributions, (83)(84) enables us to compute an analytic solution of (77) if we replace u_h^n by $\nabla \times \psi_h$; then (71) is satisfied.

So we shall define $X_h^{\prime n}$ as a solution of (77) with $\nabla \times \psi_h^n$, and we will assume that $(u_h^n \text{ o } X_h^{\prime n}, v_h)$ is calculated exactly.

Theorem 7 :
Assume Ω convex, bounded, polygonal and $u \in W^{1,\infty}(\Omega \times]0, T[)^3 \cap C^o(0, T; H^2(\Omega)^3)$, $p \in C^o(0, T; H^2(\Omega))$. Let $\{u_h^n(x)\}_n$ be the solutions in V_{oh} of

$$\frac{1}{k}(u^{n+1}h, v_h) + \nu(\nabla u_h^{n+1}, \nabla v_h) = (f^n, v_h) + \frac{1}{k}(u_h^n o X_h^{\prime n}, v_h), \quad \forall v_h \in V_{oh} \quad (88)$$

where $X_h^{\prime n}$ is calculated by (77) with u_h^n replaced by $\nabla \times \psi_h^n$ and ψ_h^n solution of (86), (87).

Then if u is the exact solution of the incompressible Navier-Stokes equations, (1)(4) with $u_\Gamma = 0$, we have

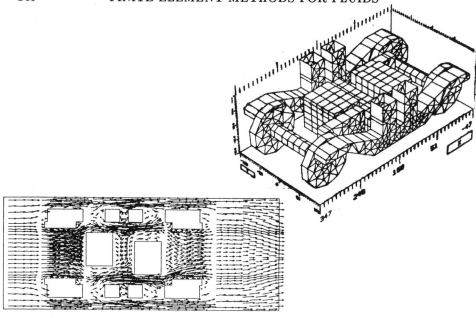

Figure 5.7 : *3D flow below a train between the wheel axis; computed with the P1 bubble/P1 element at Reynolds 1000. The figures show the trace of the mesh on the solid boundaries and the velocities in a plan section parallel to the track. (Computed by C. Parès).*

$$(|u_h^n - u(.,nk)|_0^2 + \nu k|\nabla(u_h^n - u(.,nk)|_0^2)^{\frac{1}{2}} \leq C[\frac{h^2}{k} + h + k] \qquad (89)$$

where the constant C is independent of ν, affine in $||u||_2$, $||p||_2$ and exponential in $||u||_{1,\infty}$.

Remark 1 :

Note that (89) is true even with $\nu \to 0$ (Euler's equation) but it is necessary that the solution of (1)-(4) remain in $W^{1,\infty}$ when $\nu \to 0$ (which assumes that f depends on ν). Numerically, we observe that the method remains stable even with $\nu = 0$ so long as the solution of the continuous problem is regular.

Remark 2 :

The hypothesis 'Ω convex' can be relaxed. It is only necessary in order that $(\psi \to |\nabla \times \psi|^2 + |\nabla.\psi|^2)^{1/2}$ be a norm.

To prove (89), we shall follow these three steps :
- estimate $\nabla \times \psi_h^n$ - u_h^n
- estimate X_h^n - X^n
- estimate u_h^n - u^n where u^n is the value of the exact solution at time nk.
Before starting, we note that from (1)-(4), we can deduce

$$\frac{1}{k}u^{n+1} - \nu\Delta u^{n+1} + \nabla p^{n+1} = f^n + \frac{1}{k}u^n oX^n + k|u_{,tt}|_\infty 0(1), \quad \nabla.u^{n+1} = 0 \quad (90)$$

$$u^0 \text{ given}, \ u^{n+1}|_\Gamma = 0 \qquad (91)$$

where $0(1)$ is a function bounded independently of k and ν.

Lemma 1:

$$|X_h^n - X^n|_0 \le ke^k|\nabla u^n|_\infty|\nabla \times \psi_h^n - u^n|_0 \qquad (92)$$

Proof :
From the definitions of X_h and X, we have

$$\frac{d}{d\tau}(X_h - X) = \nabla \times \psi_h(X_h) - u(x) = \nabla \times \psi_h(X_h) - u(x_h) + u(x_h) - u(x) \quad (93)$$

and so

$$\frac{d}{d\tau}|X_h - X|_0(\tau) \le |(\nabla \times \psi_h - u)oX_h|_0 + |u(x_h) - u(x)|_o \qquad (94)$$

$$\le |\nabla \times \psi_h - u|_o + |\nabla u|_\infty|X_h - X|_o(\tau)$$

Since $X_h(t) = X(t) = x$, we deduce (92) from the Bellman-Gronwall inequality (by the integration of (94) on the interval $]nk,(n+1)k[$.

Lemma 2 :

$$|\nabla \times \psi_h - u|_0 \le C|u_h^n - u^n|_0 + h|\nabla u^n|_0 \qquad (95)$$

Proof:
Let ψ^n be a solution of

$$(\nabla \times \psi^n, \nabla \times w) + (\nabla.\psi^n, \nabla.w) = (u^n, \nabla \times w), \forall w \in W \qquad (96)$$

$$\psi^n \in W = \{w \in H^1(\Omega)^n : w \times n|_\Gamma = 0, \int_{\Gamma i} w.nd\gamma = 0\} \qquad (97)$$

By subtracting (96) from (86), we see

$$(\nabla \times (\psi_h^n - \psi^n), \nabla \times w_h) + (\nabla.(\psi_h^n - \psi^n), \nabla.w_h) = (u_h^n - u^n, \nabla \times w_h) \quad (98)$$

Let ξ_h be a projection of ψ^n in W_h, that is, the solution in W_h of

$$(\nabla \times \xi_h, \nabla \times w_h) + (\nabla.\xi_h, \nabla.w_h) = (\nabla \times \psi^n, \nabla \times w_h) + (\nabla.\psi^n, \nabla.w_h), \forall w_h \in W_h \tag{99}$$

By taking $w_h = \xi_h - \psi_h^n$ in (99) and by subtracting (86), we see that

$$|\nabla \times (\xi_h - \psi_h^n|_0 \leq C|u_h - u^n|_0 \tag{100}$$

Finally, we obtain the result because u is equal to $\nabla \times \psi$ and

$$|\nabla \times (\psi_h^n - \psi)|_0 \leq |\nabla \times (\psi_h^n - \xi_h)|_0 + |\nabla \times (\xi_h - \psi)|_0 \leq C(|u_h - u|_0 + h|\nabla u^n|_0) \tag{101}$$

the last inequality comes from (100) and from the error estimate between ξ_h and ψ (the error of projection is smaller than the interpolation error).

To have existence and uniqueness for (86)+(99), the bilinear form should be coercive, which we can prove only for convex polygonal domains (Girault et al [93]).

Corollary 1 :

$$|X_h^n - X^n|_0 \leq kC(|\nabla u|_\infty)[|u_h^n - u^n|_0 + h] \tag{102}$$

Proof of Theorem 7 :

We subtract (90) from the variational form of (88) :

$$(u_h^{n+1} - u^{n+1}, v_h) + \nu k(\nabla(u_h^{n+1} - u^{n+1}), \nabla v_h) + k(\nabla(p_h^{n+1} - p^{n+1}), v_h) \tag{103}$$

$$= (u_h^n o X_h^n - u^n o X^n, v_h) + k^2 |u_{,t}|_\infty (0(1), v_h)$$

Then we proceed as for the error analysis of the Stokes problem (cf. (4.74)-(4.78)). From (103), we deduce that for all $w_h \in J_{oh}$ and all $q_h \in Q_h$, we have

$$(u_h^{n+1} - w_h, v_h) + \nu k(\nabla(u_h^{n+1} - w_h), \nabla v_h) + k(\nabla(p_h^{n+1} - q_h), v_h) \tag{104}$$

$$= (u^{n+1} - w_h, v_h) + \nu k(\nabla(u^{n+1} - w_h), \nabla v_h) + k(\nabla(p^{n+1} - q_h), v_h)$$

$$+ (u_h^n o X_h^n - u^n o X^n, v_h) + k^2 |u_{,t}|_\infty (0(1), v_h), \forall v_h \in J_{oh}$$

So (see (74) for the definition of $||.||_\nu$)

$$||u_h^{n+1} - w_h||_\nu \leq ||u^{n+1} - w_h||_\nu + k|\nabla(p^{n+1} - q_h)|_0 + |u_h^n o X_h^n - u^n o X^n|_0 + k^2 |u_{,t}|_\infty \tag{105}$$

which gives if we choose q_h equal to the interpolation of p^{n+1}

$$||u_h^{n+1} - u^{n+1}||_\nu \leq ||u_h^{n+1} - w_h||_\nu + ||w_h - u^{n+1}||_\nu \leq 2||u^{n+1} - w_h||_\nu + k||p^{n+1}||_2$$

$$\text{(106)}$$

$$+|u_h^n o X_h - u^n o X^n|_0 + k^2|u_{,t}|_\infty$$

Taking w_h as equal to the interpolation of u^{n+1}, we see that

$$||u_h^{n+1} - u^{n+1}||_\nu \leq C[(h^2 + \nu hk)||u^{n+1}||_2 \qquad \text{(107)}$$

$$+hk||p^{n+1}||_2 + k^2|u_{,t}|_\infty + |u_h^n o X_h^n - u^n o X^n|_0$$

it remains to estimate the last term ;

$$|u_h^n o X_h^n - u^n o X^n|_0 \leq |u_h^n o X_h^n - u^n o X_h^n|_0 + u^n o(X_h^n - X^n)|_0 \qquad \text{(108)}$$

$$\leq |u_h^n - u^n|_0 0 + |\nabla u^n|_\infty |X_h^n - X^n|_0 \leq (1 + Ck)||u_h^n - u^n||_\nu + C'kh$$

where C and C' depend upon $|\nabla u|_\infty$ (cf. (102)). By using a recurrence argument, the proof is completed because (107) and (108) give

$$||u_h^{n+1} - u^{n+1}||_\nu \leq (1 + C_1 k)||u_h^n - u^n||_\nu + C_2(h^2 + hk + k^2) \qquad \text{(109)}$$

6. OTHER METHODS.

In practice, all the methods given in chapter 3 for the convection-diffusion equation can be extended to the Navier-Stokes equations. Paragraph 5 is an example of such an extension. One can also use the following :

- Lax-Wendroff artificial viscosity method : however it cannot be completely explicit (same problem as in §4.1) unless one adds p_t in the divergence equation (Temam [228], Kawahara [131]);
- the upwinding method by discontinuity ; this could be superimposed with the other methods of §4.1 and of §4.2. In this way we can obtain methods which converge for all ν (cf. Fortin-Thomasset [83], for an example of this type);
- streamline upwinding method (SUPG).

6.1 SUPG and AIE (Adaptive Implicit/Explicit scheme)

As in paragraph 4.3 in chapter 3 a *Petrov-Galerkin* variational formulation for the Navier-Stokes is used with the test functions $v_h + \tau u_h \nabla v_h$ instead of v_h :

$$(u_{h,t} + u_h \nabla u_h + \nabla p_h, v_h + \tau u_h \nabla v_h) + \nu(\nabla u_h, \nabla v_h) - \nu \sum_l \int_{T_l} \tau u_h \nabla v_h \Delta u_h$$

$$= (f, v_h + \tau u_h \nabla v_h) \quad \forall v_h \in V_{0h} (\nabla.u_h, q_h) = 0 \quad \forall q_h \in Q_h.$$

Here τ is a parameter of order h, T_l is an element; we have used the simplest SUPG method where the viscosity is added in space only. Johnson [125] rightly suggest in their error analysis the use of space-time discretisation (see 3.4.3) .

A semi-implicit time discretisation gives the following scheme:

$$(\frac{u_h^{n+1} - u_h^n}{k} + u_h^n \nabla u_h^{n+1}, v_h + \tau u_h^n \nabla v_h) + (\nabla p_h^{n+1}, v_h) + (\nabla p_h^n, \tau u_h^n \nabla v_h)$$

$$+\nu(\nabla u_h^{n+1}, \nabla v_h) - \nu \sum_l \int_{T_l} \tau u_h^n \nabla v_h \Delta u_h^n = (f^{n+1}, v_h + \tau u_h^n \nabla v_h) \quad \forall v_h \in V_{0h}$$

$$(\nabla.u_h^{n+1}, q_h) = 0 \quad \forall q_h \in Q_h.$$

As noted in Tezduyar [227], it is possible to choose τ so as to have symmetric linear systems because the non-symmetric part comes from:

$$(\frac{u_h^{n+1}}{k}, \tau u_h^n \nabla v_h) + (u_h^n \nabla u_h^{n+1}, v_h)$$

Now this would be zero if $\tau = k$ and if $u_h^n \nabla u_h^{n+1}$ was equal to $\nabla.(u_h^n \otimes u_h^{n+1})$. So this suggests the following modified scheme:
Find $u_h^{n+1} - u_\Gamma \in V_{0h}$ and $p_h^{n+1} \in Q_h$ such that

$$(\frac{u_h^{n+1} - u_h^n}{k} + u_h^n \nabla u_h^{n+1}, v_h + k u_h^n \nabla v_h) + (\nabla.u_h^n, u_h^{n+1}.v_h)$$

$$+(\nabla p_h^{n+1}, v_h) + (\nabla p_h^n, k u_h^n \nabla v_h) + \nu(\nabla u_h^{n+1}, \nabla v_h) - \nu \sum_l \int_{T_l} k u_h^n \nabla v_h \Delta u_h^n$$

$$= (f^{n+1}, v_h + k u_h^n \nabla v_h) \quad \forall v_h \in V_{0h}$$

$$(\nabla.u_h^{n+1}, q_h) = 0 \quad \forall q_h \in Q_h.$$

Naturally a Crank-Nicolson $O(k^2)$ scheme can be derived in the same way. Similarly a semi-explicit first order scheme could be:

$$(\frac{u_h^{n+1} - u_h^n}{k} + u_h^n \nabla u_h^n, v_h + k u_h^n \nabla v_h) + (\nabla p_h^{n+1}, v_h) + (\nabla p_h^n, k u_h^n \nabla v_h)$$

$$+\nu(\nabla u_h^n, \nabla v_h) - \nu \sum_l \int_{T_l} k u_h^n \nabla v_h \Delta u_h^n = (f^n, v_h + k u_h^n \nabla v_h) \quad \forall v_h \in V_{0h}$$

$$(\nabla.u_h^{n+1}, q_h) = 0 \quad \forall q_h \in Q_h.$$

Still following Tezduyar [227] to improve the computing time we can use the explicit scheme in regions where the local Courant number is large and the implicit scheme when it is small; the decision is made element by element based on the local Courant number for convection, $|u_h^n|_{T_l} k/h_l$ and for diffusion $\nu k/h_l^2$, and on a measure of the gradient per element size of u_h^n; here h_l is the average size of element l. So we define n numbers by

$$\beta_i^l(c) = \max\{|u_h^n|_{T_l}\frac{k}{h_l} - 1, (\frac{U_i^l - u_i^l}{U_i^\Omega - u_i^\Omega})\sum_j |\frac{\partial u_{hi}}{\partial x_j}| - c\}$$

where U_i^l (resp u_i^l) is the maximum (resp minimum) of $u_{hi}(x)$ on element T_l and similarly for U_i^Ω, u_i^Ω on Ω instead of T_l.

The constant c is chosen for each geometry; then if $\beta_i^l(c) > 0$ the contribution to the linear system of the convection terms on element l in the $i - th$ momentum equation is taken explicit (with u_h^n instead of u_h^{n+1}) and otherwise implicit. Similarly if $\nu k/h_l^2$ is less than one, the contribution to the matrix of the linear system from $\nu \int_{T_l} \nabla u_h, \nabla v_h$ is taken with $u_h = u_h^n$ (explicit) and with u_h^{n+1} otherwise.

7. TURBULENT FLOW SIMULATIONS.

7.1. Reynolds' Stress Tensor.

As we have said in the beginning of the chapter, there are reasons to believe that when $t \to \infty$, the solution of (1)-(4) evolves in a space of dimension proportional to $\nu^{-9/4}$.

For practical applications ν (= 1/Re) is extremely small and so it is necessary to have a large number of points to capture the limiting solutions in time.

We formulate the following problem (the Reynolds problem) :
Let u_w^ν be a (random) solution of

$$u_{,t} + u\nabla u + \nabla p - \nu \Delta u = 0, \quad \nabla.u = 0 \text{ in } \Omega \times]0, T[\tag{110}$$

$$u(x,0) = u^o(x) + w(x,\omega), \quad u|_\Gamma = u_\Gamma \tag{111}$$

where $w(x,.)$ is a random variable having zero average.

Let $< >$ be the average operator with respect to the law of u introduced by w. Can we calculate $< u >, < u \otimes u > ...$?

This problem corresponds closely to the goals of numerical simulation of the Navier-Stokes equations at large Reynolds number because, when $\nu << 1$, u is unstable with respect to initial conditions and so the details of the flow cannot be reproduced from one trial to the next. Thus, it is more interesting to find $< u >$. One may also be interested in $< u^2 >$ and $\nu < |\nabla u|^2 >$.

Only heuristic solutions of the Reynolds problem are known (see Lesieur [150] for example) but let us do the following reasoning :

If we continue to denote by u the average $< u_w^\nu >$ and by u' the difference $u_w^\nu - < u_w^\nu >$, then (110) becomes :

$$u_{,t} + u\nabla u + \nabla p - \nu\Delta u + \nabla.u' \otimes u' = -(u'_{,t} + u'\nabla u + u\nabla u' + \nabla p' - \nu\Delta u') \quad (112)$$

$$\nabla.u' + \nabla.u = 0 \quad (113)$$

because $u\nabla u = \nabla.(u \otimes u)$ when $\nabla.u = 0$.

If we apply the operator $< >$ to (112) and (113), we see that

$$u_{,t} + u\nabla u + \nabla p - \nu\Delta u + \nabla. < u' \otimes u' >= 0, \nabla.u = 0 \quad (114)$$

which is the *Reynolds equation* and

$$R =< u' \otimes u' > \quad (115)$$

is the *Reynolds tensor*. As it is not possible to find an equation for R as a function of u, we will use an hypothesis (called a closure assumption) to relate R to ∇u.

It is quite reasonable to relate R to ∇u because the turbulent zones are often in the strong gradient zones of the flow. But then $R(\nabla u)$ cannot be arbitrary chosen because we must keep (114) invariant under changes of coordinate systems. In fact, it would be absurd to propose an equation for u which is independent upon the reference frame. One can prove (Chacon-Pironneau [50]) that in this case the only form possible for R is (see also Speziale [218]).

$$R = a(|\nabla u + \nabla u^T|^2)I + b(|\nabla u + \nabla u^T|^2)(\nabla u + \nabla u^T). \quad (116)$$

in 2D. In 3D

$$R = aI + b(\nabla u + \nabla u^T) + c(\nabla u + \nabla u^T)^2 \quad (117)$$

where a, b and c are functions of the 2 nontrivial invariants of $(\nabla u + \nabla u^T)$. So we get

$$\nabla.R = \nabla a + \nabla.[b(\nabla u + \nabla u^T)] + \nabla.[c(\nabla u + \nabla u^T)^2] \quad (118)$$

but ∇a is absorbed by the pressure (we change p to $p + a$) and so a law of the type

$$R = b(\nabla u + \nabla u^T) + c(\nabla u + \nabla u^T)^2 \quad (119)$$

has the same effect. In 2D, it is enough to specify a function in one variable $b(s)$ and in 3D two functions in two variables $b(s, s'), c(s, s')$ where s and s' are the two invariants of $\nabla u + \nabla u^T$.

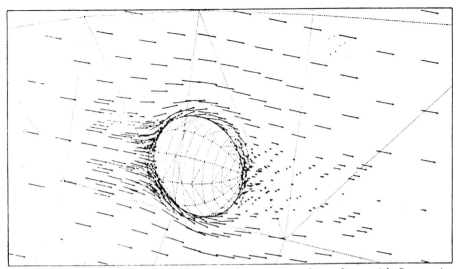

Figure 5.8 : *Turbulent flow simulation around a sphere with Smagorinsky's turbulent model (Computed by C. Parès).*

7.2. The Smagorinsky hypothesis [216] :

Smagorinsky proposes that $b = ch^2|\nabla u + \nabla u^T|$, that is

$$R = -ch^2|\nabla u + \nabla u^T|(\nabla u + \nabla u^T), \quad c \cong 0.01 \tag{120}$$

where h is the average mesh size used for (114). This hypothesis is compatible with the symmetry and a bidimensional analysis of R ; it is reasonable in 2D but not sufficient in 3D (Speziale [219]). The fact that h is involved is justified by an ergodic hypothesis which amounts to the identification of $< >$ with an average operator in a space on a ball of radius h. Numerical experiments show that we obtain satisfying results (Moin-Kim [173]) when we have sufficient points to cause u' to correspond to the beginning of the inertial range of Kolmogorov.

7.3. The $k - \epsilon$ hypothesis (Launder-Spalding [143]):

The so-called $k-\epsilon$ model was introduced and studied by Launder-Spalding [143], Rodi [200] among others.

Define the kinetic energy of the turbulence k and average rate of dissipation of energy of the turbulence ϵ by:

$$k = \frac{1}{2} < |u'|^2 > \tag{121}$$

$$\epsilon = \frac{\nu}{2} < |\nabla u' + \nabla u'^T|^2 >; \tag{122}$$

then R, k, ϵ are modeled by

$$R = \frac{2}{3}kI - c_\mu \frac{k^2}{\epsilon}(\nabla u + \nabla u^T), \quad (c_\mu = 0.09) \tag{123}$$

$$k_{,t} + u\nabla k - \frac{c_\mu}{2}\frac{k^2}{\epsilon}|\nabla u + \nabla u^T|^2 - \nabla.(c_\mu \frac{k^2}{\epsilon}\nabla k) + \epsilon = 0 \tag{124}$$

$$\epsilon_{,t} + u\nabla\epsilon - \frac{c_1}{2k}|\nabla u + \nabla u^T|^2 - \nabla.(c_\epsilon \frac{k^2}{\epsilon}\nabla\epsilon) + c_2\frac{\epsilon^2}{k} = 0 \tag{125}$$

with $c_1 = 0.1256$, $c_2 = 1.92$, $c_\epsilon = 0.07$.

A rough justification for this set of equations is as follows. First one notes that k^2/ϵ has the dimension of a length square so it makes sense to use (123).

To obtain an equation for k, (112) is multiplied by u' and averaged (we recall that $A : B = A_{ij}B_{ij}$) :

$$\frac{1}{2} <u'^2>_{,t} + <u'\otimes u'> : \nabla u + \frac{1}{2}u\nabla <u'^2> + \nabla. <p'u'> -\nu <u'\Delta u'>$$

$$+\frac{1}{2}\nabla. <u'^2 u'> = 0;$$

That is to say with (123)

$$k_{,t} - c_\mu \frac{k^2}{\epsilon}(\nabla u + \nabla u^T) : \nabla u + u\nabla k - \nu <u'\Delta u'> = - <u'\nabla\frac{u'^2}{2}> - \nabla. <p'u'>$$

The last 3 terms cannot be expressed in terms of u, k and ϵ, so they must be modelled. For the first we use an ergodicity hypothesis and replace an ensemble average by a space average on a ball of centre x and radius r, $B(x,r)$:

$$- <u'\Delta u'> = <|\nabla u'|^2> + \int_{\partial B(x,r)} u'.\frac{\partial u'}{\partial n}d\gamma.$$

By symmetry (quasi-homogeneous turbulence) the boundary integral is small.

The second term is modelled by a diffusion:

$$<u'\nabla\frac{u'^2}{2}> \cong -\nabla. <u'\otimes u'> \nabla k;$$

If u'^2 and u' were stochastically independent and if the equation for u'^2 was linear, it would be exact up to a multiplicative constant, the characteristic time of u'. The third term is treated by a similar argument as the first one so it is conjectured to be small. So the equation for k is found to be :

$$k_{,t} + u\nabla k - \frac{c_\mu}{2}\frac{k^2}{\epsilon}|\nabla u + \nabla u^T|^2 - \nabla.(c_\mu \frac{k^2}{\epsilon}\nabla k) + \epsilon = 0 \tag{126}$$

To obtain an equation for ϵ one may take the curl of (112), multiply it by $\nabla \times u'$ and use an identity for homogeneous turbulence:

$$\epsilon = \nu < |\nabla \times u'|^2 >$$

Letting $\omega' = \nabla \times u'$, one obtains:

$$0 = 2\nu < \omega'.(\omega'_{,t} + (u + u')\nabla(\omega + \omega') - (\omega + \omega')\nabla(u + u') - \nu\Delta\omega') >$$

$$\cong \epsilon_{,t} + u\nabla\epsilon + < u'\nabla\nu\omega'^2 > -2\nu < \omega'\nabla \times (u' \times \omega) >$$

$$-2\nu(< \omega' \otimes \omega' >: \nabla u + \nabla. < (\omega' \otimes \omega')u' >) + 2\nu^2 < |\nabla\omega'|^2 >$$

because

$$\nabla \times (u' \times \omega) = \omega\nabla u' - u'\nabla\omega.$$

the term $< \omega'\nabla \times (u' \times \omega) >$ is neglected for symmetry reasons; $< u'\nabla\nu\omega'^2 >$ is modelled by a diffusion just as in the k equation; by frame invariance, $< \omega' \otimes \omega' >$ should also depend only on $\nabla u + \nabla u^T$, k and ϵ therefore in 2D it can only be proportional to $\nabla u + \nabla u^T$ and by a dimension argument it must be proportional to k. The third term is neglected because it has a small spatial mean and the last term is modelled by reasons of dimension by a quantity proportional to ϵ^2/k. Finally one obtains :

$$\epsilon_{,t} + u\nabla\epsilon - \frac{c_1}{2k}|\nabla u + \nabla u^T|^2 - \nabla.(c_\epsilon\frac{k^2}{\epsilon}\nabla\epsilon) + c_2\frac{\epsilon^2}{k} = 0 \qquad (127)$$

The constants are adjusted so that the model makes good predictions for a few simple flows such as turbulence decay behind a grid.

Natural boundary conditions could be

$$k, \epsilon \text{ given at } t = 0 \text{ ; } k|_\Gamma = 0, \quad \epsilon|_\Gamma = \epsilon_\Gamma. \qquad (128)$$

however an attempt can be made to remove the boundary layers from the computational domain by considering "wall conditions" (see Viollet [235] for further details)

$$k|_\Gamma = u^{*2}c_\mu^{-\frac{1}{2}}, \quad \epsilon|_\Gamma = \frac{u^{*3}}{K\delta} \qquad (129)$$

$$u.n = 0, \quad \alpha u.\tau + \beta\frac{\partial u.\tau}{\partial n} = \gamma \qquad (130)$$

where K is the Von Karman constant ($K = 0.41$), δ the boundary layer thickness, u^* the friction velocity, $\beta = c_\mu k^2/\epsilon$, $\alpha = c_\mu k^2/[\epsilon\delta(B + log(\delta/D))]$ where D is a roughness constant and B is a constant such that $u.\tau$ matches approximately the viscous sublayer. To compute u^*, Reichard's law may be inverted (by Newton's method for example) :

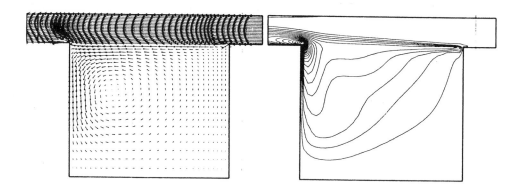

Figure 5.9 : *Computation of turbulent flow in a cavity with the $k - \epsilon$ model. The Q2/P1-discontinuous quadrilateral element has been used. The figures show the streamlines and the iso-k lines. (Computed by B. Cardot).*

$$u^*[2.5log(1 + 0.4\frac{\delta u^*}{\nu}) + 7.8(1 - e^{-\frac{\delta u^*}{11\nu}} - \frac{\delta u^*}{11\nu}e^{-0.33\frac{\delta u^*}{\nu}})] = u.\tau$$

So in reality α and β in (3a) are nonlinear functions of $u.\tau, \epsilon, k$.

Positiveness of ϵ and k.

For physical and mathematical reasons it is essential that the system yields positive values for k and ϵ. At least in some cases it is possible to argue that if the system has a smooth solution for given positive initial data and positive Dirichlet conditions on the boundaries then k and ϵ must stay positive at later times.

For this purpose we looks at

$$\theta = \frac{k}{\epsilon}.$$

If D_t denotes the total derivative operator, $\partial/\partial t + u\nabla$ and E denotes $(1/2)|\nabla u + \nabla u^T|^2$, then

$$D_t\theta = \frac{1}{\epsilon}D_t k - \frac{k}{\epsilon^2}D_t\epsilon = \theta^2 E(c_\mu - c_1) + \frac{c_\mu}{\epsilon}\nabla.(\frac{k^2}{\epsilon}\nabla k) - c_\epsilon\frac{k}{\epsilon^2}\nabla.(\frac{k^2}{\epsilon}\nabla\epsilon) - 1 + c_2$$

$$= \theta^2 E(c_\mu - c_1) - 1 + c_2 + c_\mu\nabla.k\theta\nabla\theta + 2c_\mu\nabla\theta.\nabla(\frac{\theta}{k}) + (c_\mu - c_\epsilon)\frac{\theta^2}{k}\nabla.k\theta\nabla\epsilon$$

Because $c_\mu < c_1$ and $c_2 > 1$, θ will stay positive and bounded when there are no diffusion terms because θ is a solution of a stable autonomous ODE along the streamlines:

$$D_t\theta = \theta^2 E(c_\mu - c_1) - 1 + c_2$$

Also θ cannot become negative when $c_\mu = c_\epsilon$ because the moment the minimum of θ is zero at (x,t) we will have:

$$\nabla\theta = 0, \quad \theta = 0$$

By writing the θ equation at this point, we obtain :

$$\theta_{,t} = c_2 - 1 > 0.$$

which is impossible because θ cannot become negative unless $\theta_{,t} \leq 0$.

Similarly we may rewrite the equation for k in terms of θ :

$$D_t k - \frac{c_\mu}{2}k\theta|\nabla u + \nabla u^T|^2 - \nabla.(c_\mu k\theta\nabla k) + \frac{k}{\theta} = 0$$

Here it is seen that a minimum of k with $k = 0$ is possible only if $D_t k = 0$ at that point, which means that k will not become negative.

Note however that k may have an exponential growth if $c_\mu\theta^2 |\nabla u + \nabla u^T|^2 > 2$; this will be a valuable criterion to reduce the time step size when it happens.

7.4. Numerical Methods.

Let us consider the following model :

$$u_{,t} + u\nabla u + \nabla p - \nabla.[\nu(|\nabla u + \nabla u^T|)(\nabla u + \nabla u^T)] = 0, \quad \nabla.u = 0 \qquad (131)$$

$$u.n = 0, \quad au.\tau + \nu\frac{\partial u.\tau}{\partial n} = b \qquad (132)$$

where a, b are given and ν is an increasing positive function of $|\nabla u + \nabla u^T|$.

This problem is well posed (there is even a uniqueness of solution in 3D under reasonable hypotheses on ν (cf. Lions [153])).

The variational formulation can be written in the space of functions having zero divergence and having normal trace zero :

$$(u_{,t}, v) + (u\nabla u, v) + \frac{1}{2}(\nu(|\nabla u + \nabla u^T|)(\nabla u + \nabla u^T), \nabla v + \nabla v^T) \qquad (133)$$

$$+ \int_\Gamma [au.v - bv]d\gamma = 0$$

$$\forall v \in J_{on}(\Omega); \quad u \in J_{on}(\Omega) = \{v \in H^1(\Omega)^3 : \nabla.u = 0, v.n|_\Gamma = 0\} \qquad (134)$$

By discretising the total derivative, we can consider a semi-implicit scheme,

$$\frac{1}{k}(u_h^{n+1} - u_h^n o X_h^n, v_h) + \frac{1}{2}(\nu_h^n(\nabla u_h^{n+1} + (\nabla u_h^{n+1})^T), \nabla v_h + (\nabla v_h)^T) \quad (135)$$

$$+ \int_\Gamma (au_h^n - b)v_h d\gamma = 0, \quad \forall v_h \in J_{onh}$$

$$u_h^{n+1} \in J_{onh} = \{v_h \in V_h : (\nabla.v_h, q_h) = 0 \quad \forall q_h \in Q_h; \quad v_h.n_h|_\Gamma = 0\} \quad (136)$$

where V_h and Q_h are as in chapter 4 and where :

$$\nu_h^n = \nu(|\nabla u_h^n + (\nabla u_h^n)^T|) \quad (137)$$

There is a difficulty with J_{onh} and the choice of the approximated normal n_h, especially if Ω has corners.

We can also replace J_{onh} by

$$J'_{onh} = \{v_h \in V_h : \quad (v_h, \nabla q_h) = 0 \quad \forall q_h \in Q_h\} \quad (138)$$

because

$$0 = (u, \nabla q) = -(\nabla.u, q) + \int_\Gamma u.nq d\gamma, \quad \forall q \quad \Rightarrow \quad \nabla.u = 0, \quad u.n|_\Gamma = 0. \quad (139)$$

With J'_{onh} the slip boundary conditions are satisfied in a weak sense only but the normal n_h does not now appear .

The techniques developed for Navier-Stokes equations can be adapted to this framework, in particular the solution of the linear system (134) can be carried out with the conjugate gradient algorithm developed in chapter 4. However we note that the matrices would have to be reconstructed at each iteration because ν depends on n.

To solve (114), (123) (the equations $k - \epsilon$), we can use the same method ; we add to (133)-(134) (k is replaced by q)

$$(q_h^{n+1} - q_h^n o X_h^n, w_h) + kc_\mu(\frac{q_h^{n^2}}{\epsilon_h^n}\nabla q_h^{n+1}, \nabla w_h) \quad (141)$$

$$+(\int_{nk}^{(n+1)k} [\epsilon_h^n - c_\mu\frac{q_h^{n^2}}{2\epsilon_h^n}|\nabla u_h^{n+1} + \nabla u_h^{n+1}|^2](X(t))dt, w_h) = 0, \quad \forall w_h \in W_{oh}$$

$$(\epsilon_h^{n+1} - \epsilon_h^n o X_h^n, w_h) + kc_\epsilon(\frac{q_h^{n^2}}{\epsilon_h^n}\nabla \epsilon_h^{n+1}, \nabla w_h^n) \quad (142)$$

$$+(\int_{nk}^{(n+1)k} -[\frac{c_1}{2}q_h^n|\nabla u_h^{n+1}+\nabla u_h^{n+1T}|^2 + c_2\frac{\epsilon_h^n}{q_h^n}](X(t))dt, w_h) = 0, \quad \forall w_h \in W_{oh}$$

The integrals from nk to $(n+1)k$ are carried out along the streamlines in order to stabilize the numerical method (Goussebaile-Jacomy [99]).

So, at each iteration, we must
- solve a Reichard's law at each point on the walls,
- solve (134),
- solve the linear systems (141)-(142).

The algorithm is not very stable and converges slowly in some cases but it may be modified as follows.

As in Goussebaile et al.[99] the equations for $q-\epsilon$ are solved by a multistep algorithm involving one step of convection and one step of diffusion. However in this case the convection step is performed on q, θ rather than on q, ϵ.

The equation for q is integrated as follows:

$$(q_h^{n+1}, w_h) - (q_h^n o X_h^n, w_h) + (q_h^{n+1}\int_{nk}^{(n+1)k}(\frac{c_\mu}{2}\frac{q_h^n}{\epsilon_h^n}|\nabla u_h^n + \nabla u_h^n|^2 - \frac{\epsilon_h^n}{q_h^n}), w_h) \quad (143)$$

$$+kc_\mu(\frac{q_h^{n2}}{\epsilon_h^n}\nabla q_h^{n+1}, \nabla w_h) = 0 \quad \forall w_h \in Q_{oh} = \{w_h \in Q_h : w_h|_\Gamma = 0\}$$

$$q_h - k_{\Gamma h} \in Q_{oh} \quad (144)$$

But the equation for ϵ is treated in two steps via a convection of $\theta = q/\epsilon$ that does not include the viscous terms

$$(\theta_h^{n+\frac{1}{2}}, w_h) + (\theta_h^{n+\frac{1}{2}}\int_{nk}^{(n+1)k}\frac{1}{2}\theta_h^n|\nabla u_h^n + \nabla u_h^{nT}|^2(c_1 - c_\mu), w_h) \quad (145)$$

$$= (\theta_h^n o X_h^n, w_h) + (c_2 - 1, w_h)k$$

where $\theta_h^n = q_h^n/\epsilon_h^n$. Then $\epsilon_h^{n+1/2}$ is found as

$$\epsilon_h^{n+\frac{1}{2}} = \frac{q_h^{n+1}}{\theta_h^{n+\frac{1}{2}}} \quad (146)$$

and a diffusion step can be applied to find ϵ_h^{n+1} :

$$(\epsilon_h^{n+1}, w_h) + kc_\epsilon(\frac{q_h^{n2}}{\epsilon_h^n}\nabla \epsilon_h^{n+1}, \nabla w_h) = (\epsilon_h^{n+\frac{1}{2}}, w_h) \quad (147)$$

$$\forall w_h \in Q_{oh}; \epsilon_h^{n+1} - \epsilon_{\Gamma h} \in Q_{oh}$$

Notice that this scheme cannot produce negative values for q_h^{n+1} and ϵ_h^{n+1} when

$$\int_{]x,X_h^n(x)[} (\frac{c_\mu}{2}\theta_h^n|\nabla u_h^n + \nabla u_h^n|^2 - \frac{1}{\theta_h^n}) > -1 \qquad (148)$$

because $c_1 > c_\mu$ and $c_2 > 1$ so (145) generates only positive θ while in (143) the coefficient of q_h^{n+1} is positive. Note that (148) is a stability condition when the production terms are greater than the dissipation terms.

Chapter 6

Euler, Navier-Stokes
and the shallow water equations

1. ORIENTATION

In this chapter, we have grouped together the three principal problems of fluids mechanics where the *nonlinear hyperbolic* tendency dominates.

First, we shall recall some general results on nonlinear hyperbolic equations. Then we shall give some finite element methods for the Euler equations and finally we shall extend the results to the compressible Navier-Stokes equations.

In the second part of this chapter, we shall present the necessary modifications needed to apply the previous schemes to the incompressible Navier-Stokes equations averaged in x_3 : Saint-Venant's shallow water equations.

2. COMPRESSIBLE EULER EQUATIONS

2.1. Position of the problem

The general equations of a perfect fluid in 3 dimensions can be written as (cf. (1.2), (1.7), (1.12))

$$W_{,t} + \nabla . F(W) = 0 \tag{1}$$

and, in relation to (1.12)

$$E = \rho(e + \frac{1}{2}|u|^2)$$

and

$$W = [\rho, \rho u_1, \rho u_2, \rho u_3, E]^T \tag{2}$$

$$F(W) = [F_1(W), F_2(W), F_3(W)]^T \tag{3}$$

$$F_i(W) = [\rho u_i, \rho u_1 u_i + \delta_{1i} p, \rho u_2 u_i + \delta_{2i} p, \rho u_3 u_i + \delta_{3i} p, u_i(E + p)]^T \tag{4}$$

$$p = (\gamma - 1)(E - \frac{1}{2}\rho|u|^2) \tag{5}$$

The determination of a set of Dirichlet boundary conditions for (1) to be well defined is a difficult problem. Evidently we need initial conditions

$$W(x, 0) = W^o(x) \quad \forall x \in \Omega \tag{6}$$

and to determine the boundary conditions, we write (1) as :

$$W_{,t} + \sum A_i(W)\frac{\partial W}{\partial x_i} = 0 \tag{7}$$

where $A_i(W)$ is the 5×5 matrix whose elements are the derivative of F_i with respect to W_j.

Let $n_i(x)$ be a component of the external normal to Γ at x. One can show that

$$B(W, n) = \sum A_i(W)n_i \tag{8}$$

is diagonalizable with eigenvalues

$$\lambda_1(n) = u.n - c, \quad \lambda_2(n) = \lambda_3(n) = \lambda_4(n) = u.n, \quad \lambda_5(n) = u.n + c \tag{9}$$

where $c = (\gamma p/\rho)^{1/2}$ is the velocity of sound. Finally, by analogy with the linear problem, we see that it is necessary to give Dirichlet conditions at each point x of Γ with negative eigenvalues. We will then have 0, 1, 4 or 5 conditions on the components of W according to different cases. The important cases are:

- supersonic entering flow ($u.n < 0, |u.n| > c$) : 5 conditions for example, $\rho, \rho u_i, p$
- supersonic exit flow : 0 conditions.
- subsonic entering flow : 4 conditions, independent combinations of $W.l^i$ where l^i is the left eigenvector associated to the eigenvalue $\lambda_i < 0$.
- subsonic exit flow : 1 condition (in general, on pressure)
- flow slipping at the surface ($u.n = 0$): 1 condition, i.e.: $u.n = 0$.

2.2. Bidimensional test problems

a) *Flow through a canal with an obstacle in the form of an arc of circle* (Rizzi-Viviand[199])

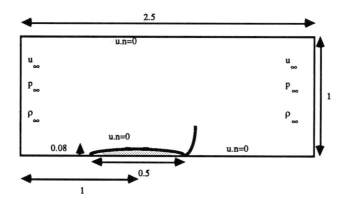

Figure 6.1 :*Stationary flow around an obstacle in the form of an arc of circle. The picture shows the geometry, the type of boundary conditions and the position of shocks.*

The height of the hump is 8% of its length. The flow is uniform and subsonic at the entrance of the canal, such that the Mach number is 0.85. We have slipping flow along the horizontal boundary and along the arc $(u.n = 0)$; the quality of the result can be seen, for example, from the distribution of entropy created by the shocks.

b) *Flow in a backward step canal* (Woodward-Colella[238])

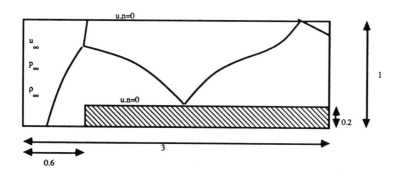

Figure 6.2 :*Shock waves configuration in the flow over a step.*

This flow is supersonic at the exit and we give only conditions at the entrance $p = 1, \rho = 1.4, u = (3,0)$. The quality of the results can be seen in the position of the shocks and of the L contact discontinuity lines.

c)_Symmetric flow with Mach 8 around a cylinder_ . (Angrand-Dervieux[2])

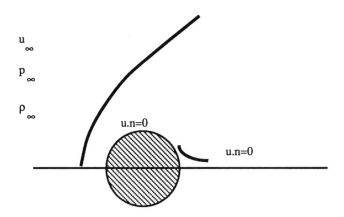

Figure 6.3:_Positions of shocks at Mach 8._

In this example, the difficulties arise because the Mach number is large, the domain is unbounded and because there is a quasi rarefaction zone behind the cylinder.

2.3. Existence

One should interpret (1) in the distribution sense because there could be discontinuities : to have uniqueness for the problem, one must add an entropy condition. The physical (reduced) entropy, s,

$$S = log\frac{p}{\rho^\gamma} \tag{10}$$

should decrease in time when we follow the fluid particles $(t \to S(X(t),t))$. In certain cases, (Kruzkov[136], Lions[153]) one can show that if the solution of (1) is perturbed by adding a viscosity $-\epsilon\Delta W$ then when $\epsilon \to 0$ it converges to the entropy solution of (1). So the existence and uniqueness of the solution of (1) can be studied as a limit of the viscosity solution.

For system (1) with the above prescribed conditions there exists the following partial results :

- an existence theorem for a local solution by a Cauchy-Kowalevska argument (cf. Lax [145])

- a global existence theorem in one dimension (cf. Majda [165])

- a global existence theorem in the case where W has only one component (scalar equation) and in arbitrary dimension (Kruzkov [136],Glimm [94], Lions [153]).

2.4. Some centered schemes

To obtain finite element methods for equations (1), we can use the method developed in chapter 3:
    - artificial viscosity methods,
    - semi-Lagrangian methods,
    - streamline viscosity methods.

a) Artificial viscosity methods (Jameson [121])
We take for (1) an explicit discretisation in time and a P^1 conforming finite element method for space :

$$\frac{1}{k}(W_h^{n+1} - W_h^n, V_h) - (F(W_h^n), \nabla V_h) + \int_\Gamma V_h.F(W_{\Gamma h}^n).n_h \qquad (11)$$

$$-(\nabla.[G(W_h^n)\nabla W_h^n], V_h) = 0 \quad \forall V_h \in U_h; \quad W_h^{n+1} - W_{\Gamma h}^n \in U_{0h}$$

where G is a viscosity tensor chosen carefully [15] and where U_{0h} is the space of vector valued functions, $piecewise - P^1$ continuous, on the triangulation and whose components corresponding to those where W_Γ is known, are zero on Γ; $W_{\Gamma h}^n$ is obtained by replacing in W_h^n the known values of W_Γ .
As viscous and convection terms are taken into account explicitly in time, there is a stability condition on k; if one is interested only in stationary solutions, mass lumping can be used to make the scheme really explicit; in addition a preconditioner can be added to accelerate it. The viscosity chosen by Jameson is approximately :

$$\nabla.(G(W)\nabla W) = -\frac{h^2}{k}(\sum \frac{\partial}{\partial x_i})[ah^3 \frac{|\Delta p|}{|p|}\frac{\partial W}{\partial z} - |b - ah^2 \frac{|\Delta p|}{|p|}| + \frac{\partial^3 W}{\partial z^3}] \qquad (12)$$

where z is the flow direction and a and b are constants of $O(1)$; however it is written as a nodal scheme in the sense that mass lumping is applied to the first term of (11) and the contribution of the diffusion term is something like

$$\sum_{j,k \in T(i)} (\sum_{m \in N(j)} (W_{.j} - W_{.m}) - \sum_{l \in N(k)} (W_{.k} - W_{.l}))$$

where $N(i)$ is the set of neighbor nodes of node i and $T(i)$ the triangles which contain the ith vertex.

b) Lax-Wendroff/Taylor-Galerkin Method (Donea [68], Lohner et al. [160])

We start with a Taylor expansion

$$W^{n+1} = W^n + kW^n_{,t} + \frac{k^2}{2}W^n_{,tt} + o(k^2) \tag{13}$$

From (1), we deduce that

$$W_{,tt} = -\nabla.(\frac{\partial F(W)}{\partial t}) = -\nabla.(F'_w(W)W_{,t}) = \nabla.(F'_w(W)\nabla.F(W)) \tag{14}$$

So (13) can be discretised as

$$(W^{n+1}_h, V_h) = (W^n_h, V_h) + k(F(W^n_h), \nabla V_h) - \frac{k^2}{2}(F'_w(W^n_h)\nabla.F(W^n_h), \nabla V_h) \tag{15}$$

$$- \int_\Gamma [kF(W^n_{\Gamma h}).n - \frac{k^2}{2}F'_w(W^n_h)\nabla.F(W^n_h)n].V_h, \quad \forall V_h \in U_{oh}$$

where U_{oh} is as in a).

Here also, one could accelerate the computation by mass lumping. In Lohner [158][160] spectacular results obtained using this method can be found.

An FCT correction (Boris-Book [34]) can be added to make the scheme more robust; if W^{pn+1} denotes the vector of values at the nodes of the solution of (15) with mass lumping in the first two terms, then the vector of values at the nodes of W^{n+1}_h is constructed from

$$W^{n+1} = W^{pn+1} + cD^* M_l^{-1}(M - M_l)W^n,$$

where c is a constant, M is the mass matrix, M_l is the lumped mass matrix and D^* is a flux limiter built by limiting the element contribution to $M_l^{-1}(M - M_l)W^n$ in such a way as to attempt to ensure that W^{n+1} is free from extrema not found in W^n or W^{pn+1}.

c) Predictor-corrector viscosity method (Dervieux [64].)

This is another form of method b. We start with a step (predictor) in which W^p_h is piecewise constant on each element T and calculated from

$$W^p_h|_T = |T|^{-1}[\int_T W^n_h - \alpha k \int_{\partial T} F(W^n_h)n_h] \tag{16}$$

with $\alpha = (1+ \sqrt{5})/2$ (cf. Lerat and Peyret [189]). Then, we compute W^{n+1}_h by solving

$$\frac{1}{k}(W^{n+1}_h - W^n_h, V_h)_h + ((1 - \beta)F(W^n_h) + \beta F(W^p_h), \nabla V_h) \tag{17}$$

$$- \int_\Gamma V_h.F(W^n_{\Gamma h})n + (G(W^n_h)\nabla W^n_h, \nabla V_h) = 0, \quad \forall V_h \in U_h$$

where $\beta = 1/2\alpha$ and where $(,)_h$ denotes the mass lumping quadrature formula; G is chosen to be proportional to h^2 and to the first derivatives of W_h^n (to give more viscosity where W is nonsmooth).

d) Streamline upwinding methods (SUPG). (Hughes-Mallet [119], Johnson-Szepessy [128])

The basic idea is to add the equations of the problem to the basis functions; we consider the scheme

$$\int_{nk}^{(n+1)k} (W_{h,t} + \nabla.F(W_h), V_h + h(V_{h,t} + A_i(W_h)V_{h,i}))dt \qquad (18)$$

$$+(W_h^{n+} - W_h^{n-}, V_h^{n-}) = 0$$

where $U^{n\pm}$ is the left or right limit of $U(t)$ when $t \to nk$. We discretise in time and in space in a coupled manner, (cf Chapter 3) for example with continuous P^1 approximation in space and P^1 discontinuous in time. Though the method is efficient numerically, (Mallet [166]) the convergence study of the algorithm suggests 3 modifications (Johnson-Szepessy [128]):

- use of entropy formulation of Euler's equations,
- replacing h by hM in (18), where M is a matrix function of W_h,
- adding another viscosity called "shock capturing" .

The final scheme is

$$\int_{nk}^{(n+1)k} (A'_o W_{h,t} + \sum_i A'_i W_{h,x_i}, V_h + h(\sum_{i\geq 0} A_i)^{-\frac{1}{2}}[A'_o V_{h,t} + \sum_i A'_i V_{h,x_i}])dt \quad (19)$$

$$+(W_h^{n+} - W_h^{n-}, V_h^{n+}) = hd \quad \forall V_h \in V_h^n$$

where

$$V_h^n = \{U_h, P^1 \text{ continuous in } x, P^1 \text{ discontinuous in } t, \qquad (20)$$

such that $U_{hi} = 0$ on the parts of Γ where W_{hi} is given $\}$

We choose $a \cong 1$, $b << 1$, $c \cong 1$, we put $\nabla U = [U_{,t}, U_{,x_i}]^T$ and we take

$$d = \int_{\Omega \times]nk,(n+1)k[} a(A'_o W_{h,t} + \sum_i A'_i W_{h,x_i}) \frac{\nabla W_h.\nabla V_h}{(b + |\nabla W_h|)} \qquad (21)$$

$$+c \int_\Omega |W_h^{n+} - W_h^{n-}|(\sum_i W_{h,x_i} V_{h,x_i})$$

The matrices A'_i are computed from the entropy function $S(W)$ (cf (10)) by $A'_o = (S_{,ww})^{-1}$, $A'_i = A_i A'_o$.

The scheme is implicit in time like the Crank-Nicolson scheme. For Burger's equation, $(F(W) = W^2/2$, W having only one component) on $\Omega = R$, result which converges towards the continuous solution when $h \to 0$ can be found in [17] ; also if W has only one component Johnson and Szepessy have shown that the method converges to the entropy solution.

In the general case one can also show that if W_h converges to W, then W satisfies the entropy condition

$$S(W)_{,t} + \sum_i S_{,w} F_i(W)_{,x_i} \leq 0$$

e) General Remarks on the above schemes

Due to the complexity of the problem, we still do not know how to prove convergence of the scheme in a general case (except maybe SUPG). So the analysis often reduces to the study of finite difference schemes on regular grids and of a linear equation (like that of chapter 3).

The schemes with artificial viscosity have the advantage of being simple but to be efficient, the regions where the viscosity plays a role must be small: this is possible with an *adaptive mesh* where we subdivide the elements if $|\nabla W|$ is large (Lohner [158], Palmério and al. [186], Bank [12], Kikuchi et al. [133]). But the stability condition forces us to reduce k if h diminishes so more iterations are needed than with implicit schemes; this requires a careful vectorization of the program and/or the use of fast quadrature formulae (Jameson [121]). Finally, the main drawback as the choice of viscosity coefficient and the time step size, one may prefer the schemes which use upwinding methods, is not being more robust (but not necessarily more accurate). On the theoretical side, it is sometimes possible to show that the schemes satisify the entropy conditions in the case of convergence (Leroux [147]).

2.5. Upwinding by discontinuity

As in chapter 3, upwinding is introduced through the discontinuities of $F(W_h)$ at the inter-element boundaries, either because W_h is a discontinuous approximation of W, or because for all W_h continuous, we know how to associate a discontinuous value at right and at left, at the inter-element boundary.

a) **General Framework** (Dervieux [64], Fezoui [78], Stoufflet et al. [222]).

As in chapter 3, for a given triangulation, we associate to each vertex q^i a cell σ^i obtained by dividing triangles (tetraedra) by the medians (by median planes). Thus we could associate to each piecewise continuous function in a triangulation a function P^o (piecewise constant) in σ^i by the formula

$$W_h^p|_{\sigma^i} = \frac{1}{|\sigma^i|} \int_{\sigma^i} W_h dx \qquad (22)$$

By multiplying (1) by a characteristic function of σ^i and by integration (Petrov-Galerkin weak formulation) we obtain, after an explicit time discretisation, the following scheme :

$$W_h^{n+1}(q^i) = W_h^n(q^i) + \frac{k}{|\sigma^i|} \int_{\partial \sigma^i} F_d(W_h^p).n \quad \forall i \qquad (23)$$

On $\sigma^i \cap \Gamma$, we take $F_d(W) = F(W)$ and we take into account the known components of W_Γ ; and elsewhere $F_d(W)$ is a piecewise constant approximation of $F(W)$ verifying

$$\int_{\partial \sigma^i} F_d(W_h^p).n = \sum_{j \neq i} \Phi(W_{h|\sigma^i}^p, W_{h|\sigma^j}^p) \int_{\partial \sigma^i \cap \sigma^j} n \qquad (24)$$

where Φ will be defined as a function of $F(W_{h|\sigma^i}^p)$ and $F(W_{h|\sigma^j}^p)$; $\Phi(u, v)$ is the numerical flux function chosen according to the qualities sought for the scheme (robustness precision ease of programming). In all cases this function should satisfy the consistency relation :

$$\Phi(V, V) = F(V), \text{ for all } V . \qquad (25)$$

b) Definition of the flux Φ :
Let $B(W,n) \in R^{5 \times 5}$ (4×4 in 2D) be such that

$$F(W).n = B(W, n).W \quad \forall W \quad \forall n \qquad (26)$$

Note that B is the same in (8) because F is homogeneous of degree 1 in W ($F(\lambda W) = \lambda F(W)$) and so B is nothing but $F_i'(W)n_i$. As we have seen that B is diagonalizable, there exists $T \in R^{5 \times 5}$ such that

$$B = T^{-1} \Lambda T \qquad (27)$$

where Λ is the diagonal matrix of eigenvalues.
We denote

$$\Lambda^\pm = diag(\pm max(\pm \lambda_i, 0)), \quad B^\pm = T^{-1} \Lambda^\pm T \qquad (28)$$

$$|B| = B^+ - B^-, \quad B = B^+ + B^- \qquad (29)$$

We can choose for Φ one of the following formulae :

$$\Phi^{SW}(V^i, V^j) = B^+(V^i)V^i + B^-(V^j)V^j \text{ (Steger-Warming)} \qquad (30)$$

$$\Phi^{VS}(V^i, V^j) = B^+(\frac{V^i + V^j}{2})V^i + B^-(\frac{V^i + V^j}{2})V^j \text{ (Vijayasundaram)} \qquad (31)$$

$$\Phi^{VL}(V^i, V^j) = \frac{1}{2}[F(V^i) + F(V^j) + |B(\frac{V^i + V^j}{2})|(V^j - V^i)] \text{ (Van Leer)} \qquad (32)$$

$$\Phi^{OS}(V^i, V^j) = \frac{1}{2}[F(V^i) + F(V^j) - \int_{V_i}^{V^j} |B(W)|dN] \text{ (0sher)} \qquad (33)$$

the guiding idea being to get $\Phi_l(V^i, V^j) \cong F(V^i)_l$ if λ_l is positive and $F(V^j)_l$ if $\lambda_l < 0$.

So Φ^{SW} , for instance, can be rewritten as follows :

$$\Phi^{SW}(V^i, V^j) = \frac{1}{2}[F(V^i) + F(V^j)] + \frac{1}{2}[|B(V^i)|V^i - |B(V^j)|V^j],$$

because

$$F(V^i) + F(V^j) = (B^+(V^i) + B^-(V^i))V^i + (B^+(V^j) + B^-(V^j))V^j;$$

The first term, if alone, would yield an upwind approximation. The second, after summation on all the neighbors of i, is an artificial viscosity term. The Van Leer and Osher schemes rely also on such an identification ; the flux of Osher is built from an integral so that it is C^1 continuous ; the path in R^5 from V^i to V^j is chosen in a precise manner along the characteristics so as to capture exactly singularities like the sonic points.

c) Integration in time.

The previous schemes are stable up to CFL of order 1 ($c|w|k/h < 1$ where c depends on the geometry); if one wants only the stationary solution, the scheme can be speeded up by a preconditioning in front of $\partial W/\partial t$; these schemes are quite robust ; 3D flows at up to Mach 20 can be computed ; but they are not precise. Schemes of order 2 are being studied.

d) Spatial approximations of order 2.

A clever way (due to Van Leer) to make the previous schemes second order is to replace V^i and V^j by the interpolates V^{i-} and V^{j+} defined on the edges where they must be computed by :

$$V^{i-} = V^i + (\nabla V)^i \frac{(q^j - q^i)}{2}$$

$$V^{j+} = V^j - (\nabla V)^j \frac{(q^j - q^i)}{2}$$

An upwinding is also introduced to compute the gradients (cf. Stoufflet et al. [222]). An alternative is to use discontinuous elements of order 2 or higher (Chavent-Jaffré [51]).

2.6. Convergence : the scalar conservation equation case

The convergence of the above scheme is an open problem except perhaps in 1D or when W has only one component (instead of 5!); this is the case of the nonlinear scalar conservation equations which govern the water-oil concentration in a porous media; let us take this as an example. Let f be in $C^1(R)$ and $\phi(x,t) \in R$ be a solution of

Figure 6.4: *Simulation of an inviscid compressible flow around a space shuttle by solving the Euler equations. (Courtesy of AMD-BA).*

$$\phi_{,t} + \nabla.f(\phi) = 0 \text{ in } Q =]0,T[\times R^n, \quad \phi(0) = \phi^o \text{ in } R^n \tag{34}$$

with ϕ^o having compact support.

One can show (Kruzkov [136]) that for $\phi^o \in L^\infty(R^n)$ there exists a unique solution (34) satisfying the entropy inequalities

$$\Phi_{c,t} + \nabla.F_c \leq 0 \text{ in } Q \text{ for all } c \tag{35}$$

where

$$\Phi_c(\phi) = |\phi - c|, \quad F_c(\phi) = (f(\phi) - f(c))sgn(\phi - c) \tag{36}$$

Moreover, ϕ is also the limit of ϕ^ϵ, $\epsilon \to 0$, the only solution of

$$\phi^\epsilon_{,t} + \nabla.(f(\phi^\epsilon)) - \epsilon\Delta\phi^\epsilon = 0 \text{ in } Q, \phi^\epsilon(0) = \phi^o \tag{37}$$

Finally, if $BV(R^n)$ denotes the space of functions of bounded variations, we have :

$$\phi^o \in L^1(R^n) \cap BV(R^n) \quad \Rightarrow \quad \phi \in L^\infty(0,T;L^1(R^n)) \cap BV(R^n) \tag{38}$$

In the finite difference world, the following definitions are introduced :

Let $\{\phi_i^n\}$ be the values at the vertices of $\phi_h^n(x)$ ($\phi_i^n = \phi_h^n(q^i)$); an explicit scheme

$$\phi_i^{n+1} = H(\{\phi_j^n\})_i \tag{39}$$

is *monotone* if H is a non-decreasing function in each of its arguments. A scheme is *TVD* (Total Variation Diminishing) if

$$\|\phi_h^{n+1}\|_{BV} \le \|\phi_h^n\|_{BV} \text{ where } \|\phi_h^n\|_{BV} = \sum_{q^i,q^j neighbors} \frac{|\phi_i - \phi_j|}{|q^i - q^j|} \qquad (40)$$

These notations can be generalized to finite element schemes at least in the case of uniform meshes. In [10] a quasi-monotone finite element scheme is constructed with upwinded fluxes and a viscosity of $O(h)$, and one gets a class of methods for which the convergence is proven.

Implicit methods with viscosity have the advantage of being automatically stable and in the case of convergence (where all the difficulties lie) convergence occurs towards an entropy solution.

For example, let us consider the following scheme :

$$\frac{1}{k}(\phi_h^{n+1} - \phi_h^n, w_h) + (\nabla.f(\phi_h^{n+1}), w_h) + \epsilon h^\alpha(\nabla\phi_h^{n+1}, \nabla w_h) = 0, \quad \forall w_h \in V_h \quad (41)$$

where V_h is a space of conforming finite elements.

By taking $w_h = \phi_h^{n+1}$, we get the stability

$$|\phi_h^{n+1}|_o^2 \le |\phi_h^n|_o |\phi_h^{n+1}|_o \qquad (42)$$

because $(\nabla.f(\phi_h), \phi_h)$ is equal to a boundary integral on Σ which is positive plus $c|\nabla\phi_h|_o^2$

Under rather strong convergence hypotheses, by taking w_h equal to the interpolation of $sgn(\phi - c)$ one can show in the limit of (41)

$$(\phi_{,t} sgn(\phi - c)) + (\nabla.f(\phi), sgn(\phi - c)) + \epsilon \lim h^\alpha(\nabla\phi_h, \nabla sgn(\phi - c)) = 0 \quad (43)$$

but

$$(\nabla\phi, \nabla sgn(\phi - c)) = \int_{\{x|\phi-c=0\}} \frac{\partial\phi}{\partial n} d\gamma > 0$$

so (43) implies ultimately (35).

As mentionned above, SUPG+shock capturing is one case where convergence can be proved by a similar argument (and many more ingredients [128])

3. COMPRESSIBLE NAVIER-STOKES EQUATIONS

3.1. Generalities

The equations (1.2), (1.7), (1.12) can also be written as a vectorial system in W. With the notations (2)-(5), we have

$$W_{,t} + \nabla.F(W) - \nabla.K(W, \nabla W) = 0 \tag{44}$$

where $K(W, \nabla W)$ is a linear 2nd order tensor in ∇W such that

$$K_{i,1} = 0, \quad K_{.,2,3,4} = \eta \nabla u + (\zeta + \frac{\eta}{3})I\nabla.u, \tag{45}$$

$$K_{.,5} = \eta(\nabla u + \nabla u^T)u + (\zeta - \frac{2}{3}\eta)u\nabla.u + \kappa\nabla(\frac{E}{\rho} - \frac{|u|^2}{2}) \tag{46}$$

The term $\nabla.K(W, \nabla W)$ is seemingly a diffusion term if $\rho, \theta > 0$ since it corresponds to $\eta\Delta u + (\eta/3 + \zeta)\nabla(\nabla.u)$ in the 2nd, 3rd and 4th equations of (44) and to $\kappa(C_v/\rho)\Delta\theta + 1/\rho C_v[|\nabla.u|^2(\zeta - 2\eta/3) + |\nabla u + \nabla u^T|\eta/2]$ in the last one (cf. (1.13)):

$$-\int_{R^3} W\nabla.K(W, \nabla W) = -\int_{R^3}[\eta\Delta u + (\frac{\eta}{3} + \zeta)\nabla(\nabla.u)]u \tag{47}$$

$$-\int_{R^3}(\kappa C_v \frac{\theta}{\rho}\Delta\theta + \frac{\theta}{\rho}C_v[|\nabla.u|^2(\zeta - 2\frac{\eta}{3}) + |\nabla u + \nabla u^T|\frac{\eta}{2}])$$

$$\geq \int_{R^3}(\eta|\nabla u|^2 + (\frac{\eta}{3} + \zeta)|\nabla.u|^2) + \kappa\frac{C_v}{\rho_{max}}\int_{R^3}|\nabla\theta|^2$$

$$+\frac{\theta_{min}}{\rho_{max}}C_v\int_{R^3}[|\nabla.u|^2(\zeta - 2\frac{\eta}{3}) + \frac{\eta}{2}|\nabla u + \nabla u^T|^2] \geq 0$$

So we can think that the explicit schemes in step 1 to integrate (1) can be applied to (44) if the stability condition on the time step k is modified :

$$k \leq Cmin[\frac{h}{|w|}, \frac{h^2}{|K|}] \tag{48}$$

where K is a function of κ, η, ζ. This modification is favorable to the schemes obtained by artificial viscosity but still when the boundary conditions are different boundary layers arise boundary layers in the neighborhood of the walls.

The boundary conditions can be treated in the strong sense (they appear in the variational space V_{oh} instead of V_h) or in the weak sense (they appear as a boundary term in the variational formulation).

Flow around a NACA0012 airfoil

A test problem is given in Bristeau et al.[43] which concerns a 2D flow around a wing profile NACA0012. Taking the origin of the reference frame at the leading edge the equation of the upper surface of the profile is :

$$y = 0.17735\sqrt{x} - 0.075597x - 0.212836x^2 + 0.17363x^3 - 0.06254x^4, \quad 0 \leq x \leq 1$$

The boundary conditions at infinity are such that the Mach number is 0.85, the Reynolds number $(u\rho/\eta)$ is 500 and the angle of attack is 0^0 or 10^0. The temperatures at infinity and on the surface are given.

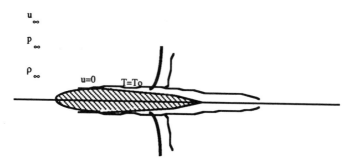

Figure 6.5: *Flow around a NACA0012. The figure shows the position of shocks and the boundary layers.*

3.2. An example

As an example, we present the results obtained by Angrand et al. [2] and Rostand [204].

The algorithm used is a modification of (16)-(17) and the boundary conditions are treated in the weak sense. The problem treated was the modeling of rarefied gas layers in the neighborhood of the surface (Knudsen layers) and so a Robin type boundary condition is applied:

$$A(W)W + K(W, \nabla W)n = g \text{ on } \Gamma \qquad (49)$$

where A is a 2nd order tensor and $g \in R^5$. (If $A \to +\infty$ we obtain the Dirichlet conditions and if $A \to 0$, the Neumann conditions).

Let U_h be the space of functions having range in R^5 continuous and piecewise P^1 on a triangulation of Ω. Problem (44) is approximated by

$$\frac{1}{k}(W_h^{n+1} - W_h^n, V_h)_h - (F_h(W_h^n), \nabla V_h) + (K_h(W_h^n, \nabla W_h^n), \nabla V_h) \qquad (50)$$

$$+ \int_{\Gamma} [F(W_h^n)V_h n + (A(W_h^n)W_h^n - g)V_h] = 0 \quad \forall V_h \in U_h$$

K_h possibly contains also the artificial viscosity if needed, F_h the upwinded flux and $(,)_h$ the mass lumping. This scheme is mostly used to calculate stationary states.

In Rostand [204], the flow around a wing profile (2D) NACA0012 has been calculated with the following boundary conditions :

$$u.n = 0, \quad n.D(u).\tau + A\rho u.\tau = 0 \quad u.D(u).n + \frac{\gamma}{Pr}\frac{\partial \theta}{\partial n} = 0 \qquad (51)$$

where $D(u) = \nabla u + \nabla u^T - 2/3 \nabla.u$ I, A is a coefficient to be chosen and Pr is the Prandtl number.

We point out the following problems which are yet to be solved :

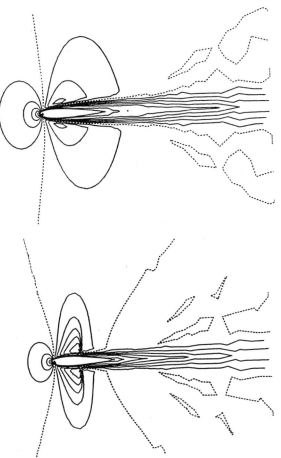

Figure 6.6: *Simulation of the compressible Navier-Stokes equations around a NACA0012 by method (60) with P^1 conforming elements. In the figure which has a shock, $A = 1$ of (61) is and in the other one $A = 100$. (Computed by Ph. Rostand).*

- find non reflecting boundary conditions at infinity towards downstream

- find a fast scheme (implicit ?) and more accurate (order 2 ?)

- remove the pressure oscillations near the walls and the leading edges in particular; it seems that one has to use a different approximation for p and u when $u << c$, which makes sense from what we know about incompressible the Navier-Stokes equations.

- make the scheme a feasible approximation for small Mach number also, (something which does not exist at the moment).

3.3 An extension of incompressible methods to the compressible case.

Since the incompressible Navier-Stokes equations studied in Chapter 5 are an approximation of the compressible equations when ρ tends to a constant, it should be possible to use the first one as an auxiliary solver for the second one.

So we shall try to associate the continuity equation and the pressure. The complete system (44) can be written as a function of $\sigma = log\rho$ as shown below, Bristeau et al.[41]. Here the method is presented with $\zeta = 0$) :

$$\sigma_{,t} + u\nabla\sigma + \nabla.u = 0$$

$$u_{,t} + u\nabla u + (\gamma - 1)(\theta\nabla\sigma + \nabla\theta) = \eta e^{-\sigma}(\Delta u + \frac{1}{3}\nabla(\nabla.u)) \tag{52}$$

$$\theta_{,t} + u\nabla\theta + (\gamma - 1)\theta\nabla.u = e^{-\sigma}(\kappa\Delta\theta + F(\nabla u)) \tag{53}$$

where $F(\nabla u) = \eta|\nabla u + \nabla u^T|^2/2 - 2\eta|\nabla.u|^2/3$ (and where θ and κ have been normalized).

An implicit time discretisation of total derivatives gives rise to a generalized Stokes system if we treat the temperature explicitly :

$$\frac{1}{k}\sigma^{n+1} + \nabla.u^{n+1} = \frac{1}{k}\sigma^n oX^n, \tag{54}$$

$$\frac{1}{k}u^{n+1} - \eta e^{-\sigma^n}(\Delta u^{n+1} + \frac{1}{3}\nabla(\nabla.u^{n+1})) + (\gamma - 1)\theta^n\nabla\sigma^{n+1}$$

$$= \frac{1}{k}u^n oX^n - (\gamma - 1)\nabla\theta^n,$$

$$(1 + \frac{1}{k}(\gamma - 1)\nabla.u^n)\theta^{n+1} - e^{-\sigma^n}\kappa\Delta\theta^{n+1} = \frac{1}{k}\theta^n oX^n + e^{-\sigma^n}F(\nabla u^n)$$

A method for solving this system is to multiply the last two equations by e^σ so as to obtain

$$\alpha\sigma + \nabla.u = g \tag{55}$$

$$\beta u - a\Delta u + b\nabla\sigma = f$$

$$\delta\theta - c\Delta\theta = h$$

where a and c are constant. So following the method of chapter 4, we take the divergence of the second equation, substitute it in the first and obtain :

$$\alpha\sigma + \nabla.[(\beta - a\Delta)^{-1}(f - b\nabla\sigma)] = 0$$

an equation that we solve by using the conjugate gradient method.

This method has the advantage of leading back to the methods studied in chapter 5 when the flow is quasi-incompressible, but has the drawback of

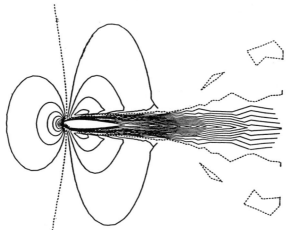

Figure 6.7: *Solution of the compressible Navier-Stokes equations around a NACA0012 by the method of paragraph 3.3 (Computed by M.O. Bristeau).*

giving a method which is non-conservative for Euler's equations when η and κ tend to 0.

4. SAINT-VENANT'S SHALLOW WATER EQUATIONS

4.1. Generalities

Let us go back to the incompressible Navier-Stokes equations with gravity $(g = -9.81)$

$$u_{,t} + \nabla.u \otimes u + \nabla p - \nu \Delta u = ge_3 \quad \nabla.u = 0 \qquad (56)$$

and assume that the domain occupied by the fluid has a small thickness, which is true in the case of lakes and seas .

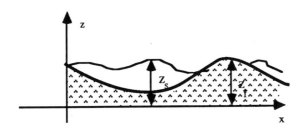

Figure 6.8 : *Vertical section through the fluid domain showing the bed·of a lake and the water surface.*

Let $z_s(x_1, x_2, t)$ be the height of the surface and $z_f(x_1, x_2, t)$ the height of the bed. The continuity equation integrated in $x_3 = z$ gives

$$\int_{z_f}^{z_s} \nabla.u dz = u_3(z_s) - u_3(z_f) + \nabla_{12}.(\int_{z_f}^{z_s} u dz) = 0 \qquad (57)$$

but by the definition of u_3 we have $dz_s/dt = u_3(z_s)$ so, taking into account the no-slip condition at the bottom and defining v, z by

$$\int_{z_f}^{z_s} u dz = (z_s - z_f)v(x_1, x_2, t), \quad z = z_s - z_f \qquad (58)$$

one can rewrite (57) as

$$z_{,t} + \nabla.(zv) = 0 \qquad (59)$$

In (56)(a) we neglect all the terms in u_3, so the third equation gives

$$\frac{\partial p}{\partial z} = g \qquad (60)$$

By putting p calculated from (60) in the two first equations of (56)(a) integrated in x_3 we see

$$(zv)_{,t} + \nabla.(zv \otimes v) + gz\nabla z - \nu\Delta(zv) \cong 0 \qquad (61)$$

The system (59), (61) constitutes the shallow water equations of Saint-Venant (for more details, see Benqué et al. [22], for example). For large regions of water (seas), the Coriolis forces, proportional to $\omega \times v$ where ω is the rotation vector of the earth must be added. Other source terms, f, in the right hand side of (61) could come from
   - the modeling of the wind effect (f constant)
   - the modeling of the friction at the bottom z_f ($f = c|v|v/z$).

If we change scales in (59), (61) as in (1.39),we obtain

$$z'_{,t'} + \nabla'.(z'v') = 0 \qquad (62)$$

$$(z'v')_{,t'} + \nabla'.(z'v' \otimes v') + F^{-2}z'\nabla'z' - R^{-1}\Delta'(z'v') = 0 \qquad (63)$$

with $R = |v|L/\nu$ and the Froude number

$$F = \frac{|v|}{\sqrt{gz}} \qquad (64)$$

If we expand (61) and divide by z,we get

$$v_{,t} + v\nabla v + g\nabla z - \nu z^{-1}\Delta(zv) = 0 \qquad (65)$$

If $F << 1$ and $R << 1$, then $g\nabla z$ dominates the convection term $v\nabla v$ and the diffusion term $\nu z^{-1}\Delta zv$; we can approximate (65) (59) by,

$$z_{,t} + \nabla.(zv) = 0 \quad v_{,t} + g\nabla z = 0 \qquad (66)$$

which is a nonlinear hyperbolic system with a propagation velocity equal to $\sqrt{gz}$.

If R $>>$ 1 so that we can neglect the viscosity term, we remark that (59), (61) is similar to the Euler equations with $\gamma = 2$, $\rho = z$ and an adiabatic approximation $p/\rho^\gamma = $ constant ; on the other hand if in (65) we set $u = zv$ then it becomes similar to the incompressible Navier-Stokes equations with an artificial compressibility such as studied by Temam [228]:

$$u_{,t} + \nabla.(z^{-1}u \otimes u) + \frac{g}{2}\nabla z^2 - \nu\Delta(u) = 0 \quad z_{,t} + \nabla.u = 0;$$

this problem is well posed with initial data on u, z and Dirichlet boundary data on u only.

But if we make the following approximation:

$$z^{-1}\Delta(zv) \cong \Delta v,$$

then (59)(61) becomes similar to the compressible Navier-Stokes equations with $\gamma = 1$:

$$v_{,t} + v\nabla v + g\nabla z - \nu\Delta v = 0$$

$$z_{,t} + \nabla.(zv) = 0;$$

An existence theorem of the type found by Matsumura-Nishida [169] may be obtained for these equations ; this means that the problem (59), (61) is well posed with the following boundary conditions :

$$z, v \text{ given at } t = 0 \qquad (67a)$$

$$v|_\Gamma \text{ given } \Gamma \text{ and } z \text{ given on all points of } \Gamma \text{ where } v.n < 0 \qquad (67b)$$

with the condition that z stays strictly positive. This fact is confirmed by an analysis of the following problem (Pironneau et al [192]):

$$g\nabla z - \nu\Delta v = 0 \quad \nabla.(zv) = 0;$$

with (67b). This reminds us that certain seemingly innocent approximations have a dramatic effect on the mathematical properties of such systems.

Boundary conditions for these shallow water equations is a serious problem; here are some difficult examples found in practical problems:
- on the banks where $z \to 0$
- in deep sea where one needs non reflecting boundary conditions if the computational domain is not to be the wholesurface of the earth.
- if the water level goes down and islands appear ($z \equiv 0$).

Test Problem :
A simple non stationary test problem is to study a wave which is axisymmetric and Gaussian in form with 2m height in a square of 5m depth and 10m side returning to rest.

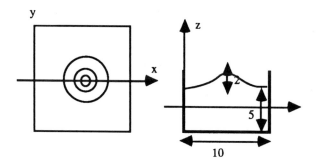

Figure 6.9 : *Computational domain (left) and vertical section through the physical domain* .

4.2. Numerical scheme in height-velocity formulation

When the Froude number is small, we shall use the methods applicable to incompressible Navier-Stokes equations rather than the methods related to Euler's equations.

We shall neglect the product term $\nabla z \, \nabla v$ (small compared to $v \nabla v$) in the development of $\Delta(zv)$ in the formulation (59), (65), that is we shall consider the system

$$z_{,t} + v \nabla z + z \nabla . v = 0 \tag{68}$$

$$v_{,t} + v \nabla v + g \nabla z - \nu \Delta v = 0 \tag{69}$$

$$z(0) = z^o, v(0) = v^o \text{ in } \Omega; \quad v|_\Gamma = v_\Gamma, \quad z|_\Sigma = z_\Gamma \tag{70}$$

For clarity, we assume that $v_\Gamma = 0$. Recall that Σ is that part of Γ where the flux enters ($v.n < 0$). If D/Dt is the total derivative, (68)-(69) can also be written

$$\frac{Dz}{Dt} + z \nabla . v = 0 \tag{71}$$

$$\frac{Dv}{Dt} + g \nabla z - \nu \Delta v = 0. \tag{72}$$

With the notation of chapter 3, we can discretise the system with the semi-implicit Euler scheme

$$\frac{1}{k}(z^{m+1} - z^m o X^m) + z^m \nabla . v^{m+1} = 0 \tag{73}$$

$$\frac{1}{k}(v^{m+1} - v^m o X^m) + g \nabla z^{m+1} - \nu \Delta v^{m+1} = 0 \tag{74}$$

For spatial discretisation, we shall use the methods given in chapter 4 for the Stokes problem : we choose $Q_h \cong L^2(\Omega)$ and $V_{oh} \cong H^1_o(\Omega)^n$ and we find $[z^{m+1}_h, v^{m+1}_h]$ solution of

$$\frac{1}{k}(z_h^{m+1}, \frac{q_h}{z_h^m}) + (\nabla.v_h^{m+1}, q_h) = \frac{1}{k}(z_h^m oX_h^m, \frac{q_h}{z_h^m}) \quad \forall q_h \in Q_h \qquad (75)$$

$$\frac{1}{k}(v_h^{m+1}, w_h) + \nu(\nabla v_h^{m+1}, \nabla w_h) + g(\nabla z_h^{m+1}, w_h) = \frac{1}{k}(v_h^m oX_h^m, w_h) \quad \forall w_h \in V_{oh} \qquad (76)$$

We recall that $X_h^m(x)$ is an approximation of $X(mk, x)$, which is a solution of

$$\frac{dX}{d\tau} = v_h^m(X(\tau), \tau), \quad X((m+1)k) = x, \qquad (77)$$

thus the boundary condition on $z_{h|\Sigma}$ appears in (75) in the calculation of $z_h^m oX_h^m(x)$, $x \in \Sigma$.

At each iteration in time, we have to solve a linear system of the type

$$\begin{pmatrix} A & B \\ B^T & -D \end{pmatrix} \begin{pmatrix} V \\ Z \end{pmatrix} = \begin{pmatrix} F \\ G \end{pmatrix} \qquad (78)$$

where

$$D_{ij} = \frac{1}{k} \int_\Omega \frac{q^i q^j}{z_h^m}, \quad B_{ij} = (\nabla q^i, w^j), \quad G_j = -\frac{1}{k}(z_h^m oX_h^m, \frac{q^j}{z_h^m})$$

$$F_j = \frac{1}{g}(v_h^m oX_h^m, w^j), \quad A_{ij} = \frac{1}{kg}(w^i, w^j) + \frac{\nu}{g}(\nabla w^i, \nabla w^j) \qquad (79)$$

As with the Stokes problem, the matrix of the linear systems is symmetric but here it is nonsingular whatever may be the chosen element $\{V_h, Q_h\}$. In fact, by using (76), to eliminate v_h^{m+1}, we find an equation for z_h^{m+1} :

$$(B^T A^{-1} B + D)Z = B^T A^{-1} F - G \qquad (80)$$

the matrix of this linear system is always positive definite (cf. (79)). It can be solved by the conjugate gradient method exactly as in the Stokes problem. A preconditioner can also be constructed in the same manner ; Goutal [100] suggests :

$$C = \frac{D}{kg} - \Delta_h \qquad (81)$$

where D is given in (79) and $-\Delta_h$ is a Laplacian matrix with Neumann condition (this preconditioning corresponds to a discretisation of the operator on z in the continuous case when $\nu = 0$).

Stability and convergence are open problems but the numerical performance of the method is good[13].

4.3. A numerical scheme in height-flux formulation

We put $D = zv$ so that (59)-(61) can be rewritten as

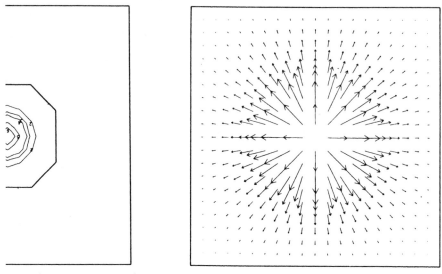

Figure 6.10: *Results for the test problem at time t = 0.2s for the shallow water equation with convection terms in the momentum equations. The computation has been done with the formulation using height and velocity, P^1 conforming elements and at Reynolds number 20; there are 400 equal triangles. The velocities and the level lines of the surface have been plotted. (Computed by V. Schoen).*

$$z_{,t} + \nabla.D = 0 \tag{82}$$

$$D_{,t} + \nabla.(\frac{1}{z}D \otimes D) + gz\nabla z - \nu\Delta D = 0 \tag{83}$$

Let us take as a model problem the case where

$$z(0) = z^o, D(0) = D^o \text{ in } \Omega \tag{84}$$

$$D = D_\Gamma \text{ on } \Gamma\times]0, T[\tag{85}$$

In [13] the following scheme is studied : with the same finite element space V_{0h} and Q_{0h}, the functions in the space Q_h which are zero on Γ, one solves

$$\frac{1}{k}(z_h^{m+1} - z_h^m, q_h) + (\nabla.D_h^{m+1}, q_h) = 0 \quad \forall q_h \in Q_h \tag{86}$$

$$\frac{1}{kg}(\frac{D_h^{m+1}}{z_h^m} - v_h^m o X_h^m, w_h) + (\nabla z_h^{m+1}, w_h) + \frac{\nu}{g}(\nabla D_h^{m+1}, \nabla\frac{w_h}{z_h^m}) \tag{87}$$

$$+\frac{1}{g}(v_h^m\nabla.v_h^m, w_h) = 0 \quad \forall w_h \in V_{oh}$$

where $v^m = D_h^m/z_h^m$ and where X_h^m is as in the previous paragraph, that is, it is calculated with $v_h : X_h^m(x) \cong x - v_h^m(x)k$.

Each iteration requires the resolution of a linear system of the type (78) but with

$$D_{ij} = \frac{1}{k}(q^i, q^j) \quad A_{ij} = \frac{1}{kg}\int_\Omega [\frac{w^i w^j}{z_h^m} + \nu \nabla w^i \nabla \frac{w^j}{z_h^m}]$$ (88)

We note that A is no longer symmetric except when $\nu = 0$. The uniqueness of the solution of (86)-(87) is no longer guaranteed except when $\nu \ll 1$ which makes it symmetric again. When $\nu \neq 0$, we can also use the following approximation

$$(\nabla \frac{D_h^{m+1}}{z_h^m}, \nabla w_h) \cong \int_\Omega \frac{1}{z_h^m}\nabla D_h^{m+1}\nabla w_h - \int_\Omega \frac{D_h^m}{(z_h^m)^2}\nabla z_h^m \nabla w_h$$ (89)

Another method can be devised by working with $z^* = z^2$ instead of z and by replacing $(z_h^{m+1} - z_h^m, q_h)$ in (86) by $(z_h^{*m+1} - z_h^{*m}, q_h/(2z_h^m))$. The convergence of this algorithm is also an open problem.

4.4. Comparison of the two schemes

The main difference between the height-velocity and height-flux formulations is in the treatment of the convection term in the continuity equation and in the boundary conditions. We expect the first formulation to be more stable for flows with high velocity. The other factors which should be considered in selecting a method are :
- the boundary conditions,
- the choice of variable to be conserved
- the presence of shocks (torrential regime, $\nu = 0$) ; (86) is in conservative form whereas (75) is not.

5. CONCLUSION

In this chapter we have briefly surveyed some recent developments in the field of compressible fluid dynamics and the shallow water equations. We have omitted reacting flows (fluid mechanics + chemistry), the Rayleigh Benard problem and flows with free surfaces because a whole book would have been necessary in place of just one short chapter; these subjects are also evolving quite rapidly so it is difficult to write on them without becoming rapidly obsolete; even for compressible flows interactions between the viscous terms and the hyperbolic terms is not well understood at the time of writing.

The architectures of computers are also changing fast; dedicated hardware for fluid mechanic problems is already appearing on the market and this are likely to influence deeply the numerical algorithms.

Finally compressible turbulence is, with hypersonic combustion, one of the challenge for the 90's.

References

[1] F. Angrand: Viscous perturbation for the compressible Euler equations, application to the numerical simulation of compressible viscous flows. M.O. Bristeau et al. ed. *Numerical simulation of compressible Navier-Stokes flows* . Notes on Numerical fluid mechanics, **18** , Vieweg, 1987.

[2] F. Angrand, A. Dervieux: Some explicit triangular finite elements schemes for the Euler equations. *J. for Num. Meth. in Fluids* , **4** , 749-764, 1984.

[3] H. Akay, A. Ecer: Compressible viscous flows with Clebsch variables. in *Numerical methods in laminar and turbulent flows* . C. Taylor, W. Habashi, M. Hafez ed. Pineridge press 1989-2001, 1987.

[4] J.D. Anderson, Jr.: Modern compressible flow with historical perspective. McGraw Hill. 1982. see also *Fundamental Aerodynamics* Mc-Graw Hill 1984.

[5] D. Arnold,F. Brezzi, M. Fortin: A stable finite element for the Stokes equations. *Calcolo* 21(**4**) 337-344 ,1984.

[6] O. Axelson, V. Barker: *Finite element solution of boundary value problems.* Academic Press 1984.

[7] K. Baba, M. Tabata: On a conservative upwind finite element schem for convection diffusion equations. *RAIRO numer Anal* . **15** ,1, 3-25, 1981.

[8] I Babuska, W. Rheinboldt: Error estimates for adaptive finite element computations. *SIAM J. Numer Comp* . **15.** 1978.

[9] I. Babuska: the p and $h - p$ version of the finite element method: the state of the art. in in *Finite element, theory and application* . D. Dwoyer, M Hussaini, R. Voigt eds. Springer ICASE/NASA LaRC series. 199-239 1988.

[10] G.K. Bachelor: *An Introduction to Fluid Dynamics.* Cambridge University Press, 1970 .

[11] A. Baker: *Finite element computational fluid mechanics* . McGraw-Hill, 1985.

[12] R. Bank: PTLMG User's guide. Dept of Math. *UCSD tech. report.* Jan 1988.

[13] R. Bank, T. Dupont, H. Yserentant: The Hierarchical basis multigrid method. *Konrad Zuse Zentrum Preprint* SC-87-1. 1987.

[14] R. Bank, L.R. Scott: On the conditioning of Finite Element equations with highly refined meshes. *Math Dept. PennState research report* 87013. 1987.

[15] R. Bank, T. Dupont, H. Yserenlant : The Hierarchichal multigrid method. *UCLA math preprint* SC87-2, 1987.

[16] C. Bardos: On the two dimensional Euler equations for incompressible fluids. *J. Math. Anal. Appl* ., **40** , 769-790, 1972.

[17] T.J. Beale, A. Majda: Rate of convergence for viscous splitting of the Navier-Stokes equations. *Math. Comp* . **37** 243-260, 1981.

[18] C. Begue, C. Conca, F. Murat, O. Pironneau: A nouveau sur les équations de Navier-Stokes avec conditions aux limites sur la pression. Note *CRAS* t 304 Serie I **1** .,23-28 , 1987

[19] C. Begue, C. Conca, F. Murat, O. Pironneau: Equations de Navier-Stokes avec conditions aux limites sur la pression. Nonlinear Partial Differential equations and their applications. Collège de France Seminar **9** , (H. Brezis and J.L. Lions ed) Pitman 1988.

[20] J.P. Benqué, O. Daubert, J. Goussebaile, H. Haugel: Splitting up techniques for computations of industrial flows. In *Vistas in applied mathematics* . A.V. Balakrishnan ed. Optimization Software inc. Springer, 1986.

[21] J.P. Benqué, A. Haugel, P.L. Viollet,: *Engineering applications of hydraulics II* . Pitman, 1982.

[22] J.P. Benque - B. Ibler - A. Keraimsi - G. Labadie: A finite element method for the Navier-Stokes eq. In Norries ed. 1110-1120, Pentech Press. 1980.

[23] J.P. Benqué, B. Ibler, A. Keramsi, G. Labadie: A new finite element method for Navier-Stokes equations coupled with a temperature equation. In T. Kawai, ed. Proc. 4th Int. Symp. on *Finite elements in flow problems* North-Holland 295-301, 1982.

[24] J.P. Benque, G. Labadie, B. Latteux: Une methode d'éléments finis pour les équations de Saint-Venant, 2nd Conf. On numerical methods in laminar and turbulent flows. Venice, 1981.

[25] M. Bercovier : Perturbation of mixed variational problems. *RAIRO série rouge* 12(**3**) 211-236 ,1978.

[26] M. Bercovier - O. Pironneau: Error Estimates for Finite Element solution of the Stokes problem in the primitive Variables. *Numer. Math* . **33,** 211-224, 1979.

[27] M. Bercovier, O. Pironneau, V. Sastri: Finite elements and characteristics for some parabolic-hyperbolic problems: *Appl. Math. Modelling* , **7** , 89-96, 1983.

[28] A. Bermudez, J. Durani: La méthode des caractéristiques pour les problèmes de convection-diffusion stationnaires. MA2N RAIRO **21** 7-26, 1987.

[29] P. Bergé, Y. Pomeau, Ch. Vidal: *De l'ordre dans le chaos* . Hermann 1984.

[30] C. Bernardi. Thèse de doctorat 3ème cycle, Université Paris 6, 1979.

[31] C. Bernardi, G. Raugel: A conforming finite element method for the time-dependant Navier-Stokes equations. *SIAM J. Numer. Anal.* **22** , 455-473, 1985.

[32] S. Boivin: A numerical method for solving the compressible Navier-Stokes equations. (to appear in *IMPACT* 1989).

[33] J. Boland, R. Nicolaides: Stability of finite elements under divergence constraints.*SIAM J. Numer. Anal* . **20** ,4, 722-731. 1983.

[34] J. Boris, D.L. Book: Flux corrected transport. *J. Comp. Phys.* **20** 397-431. 1976.

[35] A. Brandt: Multi-level adaptive solution to boundary value problems. *Math. of comp.***31**, 333-391, 1977.

[36] C. Brebbia: *The Boundary Element Method for Engineers.* Pentech, 1978.

[37] Y. Brenier: The transport-collapse algorithm, *SIAM J. Numer Anal* . **21** p1013, 1984

[38] F. Brezzi, J. Douglas: Stabilized mixed methods for the Stokes problem. (to appear).

[39] F. Brezzi: On the existence, uniqueness and approximation of saddle-point problems arising from Lagrange multipliers. *RAIRO Anal Numer.* **8** , 129-151, 1974.

[40] M.O. Bristeau, A. Dervieux: Computation of directional derivatives at the nodes: §2.4 in F. Thomasset: *Implementation of finite element methods for Navier-Stokes equations* , Springer series in comp. Physics, 48-50, 1981.

[41] M.O. Bristeau, R. Glowinski, B. Mantel, J. Periaux, C. Pouletty: Solution of the compressible Navier-Stokes equations by least-squares and finite elements. In Bristeau et al. ed:*Numerical simulation of compressible Navier-Stokes flows* . Notes on Numerical fluid mechanics, **18** , Vieweg, 1987.

[42] M.O. Bristeau , R. Glowinski, B. Mantel, J. Periaux, G. Rogé: Adaptive finite element methods for three dimensional compressible viscous flow simulation in aerospace engineering. Proc. 11th Int. Conf. on Numerical methods in fluid mechanics. Williamsburg, Virginia (1988) (*Springer* , to appear)

[43] M.O. Bristeau, R. Glowinski, J. Periaux, H. Viviand: *Numerical simulation of compressible Navier-Stokes flows* . Notes on Numerical fluid mechanics, **18** , Vieweg, 1987.

[44] M.O. Bristeau, R. Glowinski, J. Periaux, O. Pironneau, P. Perrier: On the numerical solution of nonlinear problems of fluid mech. by least squares. *Comp. Meth. in Appl. Mech ..* **17/18** March 1979.

[45] R. Brun: *Transport et relaxation dans les écoulements gazeux.* Masson 1986.

[46] L. Cafarelli, R. Kohn, L. Nirenberg: On the regularity of the solutions of the Navier-Stokes equations. *Comm. Pure and Appl. Math.* , **35** , 771-831, 1982.

[47] J. Cahouet, J.P. Chabart: Some fast 3D finite element solvers for the generalized Stokes problem. *Int. J. Numer. Methods. in Fluids* . **8**,

869-895. 1988.

[48] Cahouet, Chabart: Some fast 3-D Finite Element solvers for Generalized Stokes problem. *Rapport EDF* HE/41/87.03, 1987

[49] T. Chacon: Oscillations due to the transport of micro-structures. *SIAM J. Appl. Math.* **48** 5 p1128-1146. 1988.

[50] T. Chacon, O. Pironneau: On the mathematical foundations of the $k - \epsilon$ model. In *Vistas in applied mathematics* . A.V. Balakrishnan ed. Optimization Software inc. Springer, 1986.

[51] G. Chavent, G. Jaffré: *Mathematical methods and finite elements for reservoir simulations* . North-Holland, 1986.

[52] A.J. Chorin: Numerical study of slightly viscous flow. *J. Fluid Mech* . **57** 785-796. 1973.

[53] A. J. Chorin: A numerical Method for Solving Incompressible viscous flow problems. *J. Comp. Physics* . 2, 12-26, 1967.

[54] I. Christie, D.F. Griffiths, A.R. Mitchell, O.C. Zienkiewicz: Finite element methods for second order differential equations with significant first derivatives. *Int. J. Numer. Meth. Eng* , **10** , 1389-1396, 1976.

[55] Ph. Ciarlet: *The Finite Element Method. for Elliptic problems.* North-Holland, 1978.

[56] P.G. Ciarlet *Elasticité tridimensionnelle.* Masson Ed., 1986.

[57] B. Cockburn: The quasi-monotone schemes for scalar conservation laws. *IMA preprint 277* , University of Minnesota, Oct. 1986.

[58] G. Comte-Bellot: Ecoulement turbulent entre deux parois planes. Publications scientifiques et techniques du ministère de l'air. 1980.

[59] P. Constantin, C. Fioas, O. Manley, R. Temam: Determining modes and fractal dimension of turbulent flows.*J. Fluid Mech* . **150** , 427-440, 1985.

[60] M. Crouzeix - P.A. Raviart: Conforming and non-conforming finite element element methods for the stationary Stokes eq. *RAIRO* . R3. 33-76, 1973 .

[61] C. Cuvelier, A. Segal, A.A. van Steenhoven: *Finite Element Methods and Navier-Stokes equations.* Mathematics and its applications series, D. Reidel Publishing Co. 1986.

[62] R. Dautray, J.L. Lions: *Analyses mathématiques et calculs numériques* . Masson, 1987.

[63] J.W. Deardorff: A numerical study of 3-d turbulent channel flow at large Reynolds numbers. J. Fluid Mech. **41** , 2, 453-480, 1970.

[64] A. Dervieux: *Steady Euler simulations using unstructured meshes* . Von Karman Lecture notes series 1884-04. Mars 1985.

[65] J.A. Desideri, A. Dervieux: *Compressible Fluid Dynamics; compressible flow solvers using unstructured grids* . Von Karman lecture notes series. March 1988.

[66] Q. Dinh, R. Glowinski, J. Periaux: Domain decomposition for elliptic problems . In *Finite element in fluids* . **5** R. Gallager, J. Oden, O. Zienkiewicz, T. Kawai, M. Kawahara eds. p45-106. Wiley. 1984

[67] J.M. Dominguez, A. Bendali, S. Gallic: A variational approach to the vector potential of the Stokes problem . *J. Math. Anal. and Appl.* 107,2,

537-560, 1987.

[68] J. Donea: A Taylor-Galerkin method for convective transport problems. *J. Numer. Meth. Eng,* **20** , 101-120, 1984. (See also J. Donea, L. Quartapelle, V. Selmin: An analysis of time discretization in the Finite Element Solution of Hyperbolic problems. *J. Comp. Physics,* 70 **2** 1987.

[69] J. Douglas, T.F. Russell: Numerical methods for convection dominated diffusion problems based on combining the method of characteristics with finite element methods or finite difference method. *SIAM J. Numer Anal.* 19, **5** 871-885, 1982.

[70] J. M. Dupuy: Calculs d'écoulements potentiels transsoniques; *rapport interne* Ecole Polytechnique 1986.

[71] A. Ecer, H. Akai: Computation of steady Euler equations using finite element method. *AIAA* . **21** Nx3, p343-350. 1983.

[72] F. El-Dabaghi, J. Periaux, O. Pironneau, G. Poirier: 2d/3d finite element solution of the steady Euler equations. *Int. J. Numer. Meth. in Fluid* . **7** , p1191-1209. 1987.

[73] F. El-Dabaghi, O. Pironneau: Stream vectors in 3D aerodynamics. *Numer. Math.* , **48** 363-394, 1986.

[74] F. El-Dabaghi: Steady incompressible and compressible solution of Navier-Stokes equations by rotational correction. in *Numerical met. for fluid dynamics.* K.W.Morton and M.J. Baines. Clarendon press.p273-281. 1988.

[75] J. Essers: *Computational methods for turbulent transonic and viscous flows.* Springer, 1983.

[76] J. Feder: *Fractals* . Plenum Press. 1988.

[77] M.Festauer, J.Mandel, J.Necas: Entropy regularization of the Transonic Potential flow problem. *Commentationes Mathematicae Universitiatis Carolinae.* Prague, 1984. See also C. Necas: *transonic flow Masson* Masson 1988.

[78] F. Fezoui: Resolution des equations d'Euler par un schéma de Van Leer en éléments finis. *Rapport Inria* 358, 1985.

[79] C. Foias, R. Temam: Determination of the solution of the Navier-Stokes equations by a set of nodal values. *Math. Comput* ., **43** , 167, 117-133, 1984.

[80] M. Fortin: Calcul numérique des écoulements par la méthode des éléments. Ph.D. *Thesis* , Universite Paris 6, 1976.

[81] M. Fortin, R. Glowinski: *Augmented lagrangian methods* . North-Holland 1983.

[82] M. Fortin, A. Soulaimani: Finite element approximation of compressible viscous flows. Proc. computational methods in flow analysis, vol 2, H. Niki and M. Kawahara ed. *Okayama University Science press* . (1988)

[83] M. Fortin, F. Thomasset: Mixed finite element methods for incompressible flow problems. *J. Comp. Physics* . **31** 113-145, 1979.

[84] U. Frisch, B. Hasslacher, Y. Pomeau: Lattice-gas automata for the Navier-Stokes equation.*Physical Review letters* . **56** , 14, 1505-1508, 1986.

[85] G.P. Galbi: *Weighted residual energy methods in fluid dynamics* . Springer 1985.

[86] D. Gelder: Solution of the compressible flow equations. *Int. J. Numer. Meth. Eng.* (3) 35-43, 1987.

[87] W. Gentzsch: *Vectorization of computer programs with application to CFD* . Notes on numerical fluid mechanics. **8** Vieweg 1984.

[88] P. Geymonat, P. Leyland: Analysis of the linearized compressible Stokes problem (to appear).

[89] J.M. Ghidaglia: On the fractal dimension of attractors for viscous incompressible fluid flows. *SIAM J. Math. Anal.* , 17, 1139-1157, 1986.

[90] K. Giannakoglou, P. Chavariopoulos, K. Papaliou: Numerical computation of 2D rotational inviscid transonic flows using the decomposition of vector fields. *7th ISABE conf* . Beijing. 1985.

[91] V. Girault: Finite Element Method for the Navier-Stokes equations with non-standard boundary conditions in R^3. Rapport du Laboratoire d'Analyse Numérique de l'Université Paris 6 n^0 R87036, 1987.

[92] V.Girault, P.A. Raviart: *Finite Element Approximations of the Navier-Stokes Eq* . Lecture Notes in Math. Springer ,1979.

[93] V. Girault, P.A. Raviart: *Finite Elements methods for Navier-Stokes equations* . Springer series SCM vol 5, 1986.

[94] J. Glimm: *Comm. Pure and Appl. Math* . **18** . p697. 1965.

[95] R. Glowinski: *Numerical methods for nonlinear variationnal problems.* Springer Lectures Notes in Computationnal Physics, 1985.

[96] R. Glowinski: Le θ schéma. Dans M.O. Bristeau, R. Glowinski, J. Periaux. Numerical methods for the Navier-Stokes equations. *Comp. Phy. report* **6** , 73-187, 1987.

[97] R. Glowinski, J. Periaux, O. Pironneau: An efficient preconditioning scheme for iterative numerical solution of PDE.*Mathematical Modelling*,1979.

[98] R. Glowinski, O. Pironneau: Numerical methods for the first biharmonic problem. *SIAM Review* **21** 2 p167-212. 1979.

[99] J. Goussebaïle, A. Jacomy: Application à la thermo-hydrolique des méthodes d'éclatement d'opérateur dans le cadre éléments finis: traitement du modèle $k - \epsilon$. *Rapport EDF-LNH* HE/41/85.11, 1985.

[100] N. Goutal: Résolution des équations de Saint-Venant en régime transcritique par une méthode d'éléments finis. *Thèse* , Université Paris 6, 1987.

[101] P.M. Gresho-R.L. Sani: On pressure conditions for the incompressible Navier-Stokes equations. in *Finite Elements in Fluids* **7** . R. Gallager et al ed. Wiley 1988.

[102] P. Grisvard: *Elliptic problems in non-smooth domains* . Pitman, 1985.

[103] M. Gunzburger: Mathematical aspects of finite element methods for incompressible viscous flows. in *Finite element, theory and application* . D. Dwoyer, M Hussaini, R. Voigt eds. Springer ICASE/NASA LaRC series. 124-150 1988.

[104] L. Hackbush: *The Multigrid Method: theory and applications.* Springer series SCM, 1986.

[105] M. Hafez, W. Habashi, P. Kotiuga: conservative calculation of non isentropic transonic flows. *AIAA* . **84** 1929. 1983.

[106] L. Halpern: Approximations paraxiales et conditions absorbantes. Thèse d'état. Université Paris 6. 1986.

[107] J. Happel, H. Brenner: *Low Reynolds number hydrodynamics.* Prentice Hall 1965.

[108] F.H. Harlow, J.E. Welsh: The Marker and Cell method.*Phys. Fluids* **8** ,2182-2189, (1965).

[109] F. Hecht: A non-conforming P^1 basis with free divergence in R^3. *RAIRO serie analyse numerique* . 1983.

[110] J.C. Henrich , P.S. Huyakorn, O.C. Zienkiewicz, A.R. Mitchell: An upwind finite element scheme for the two dimensional convective equation. *Int. J. Num. Meth. Eng* . **11** , 1831-1844, 1981.

[111] J. Heywood-R. Rannacher: Finite element approximation of the nonstationnary Navier-Stokes equations (I) *SIAM J. Numer. Anal* . **19** p275. 1982.

[112] M. Holt: *Numerical methods in fluid dynamics* . Springer, 1984.

[113] P. Hood - G. Taylor: Navier-Stokes eq. using mixed interpolation. in *Finite element in flow problem* Oden ed. UAH Press,1974.

  114 K. Horiuti: Large eddy simulation of turbulent channel flow by one-equation modeling. J. Phys. Soc. of Japan. **54** , 8, 2855-2865, 1985.

[115] T.J.R. Hughes:A simple finite element scheme for developing upwind finite elements. *Int. J. Num. Meth. Eng* . **12** , 1359-1365, 1978.

[116] T. J.R. Hughes:*The finite element method.* Prentice Hall, 1987.

[117] T.J. R. Hughes, A. Brooks: A theoretical framework for Petrov-Galerkin methods with discontinuous weighting functions: application to the streamline upwind procedure. In *Finite Elements in Fluids,* R. Gallagher ed. Vol 4., Wiley, 1983.

[118] T.J.R. Hughes, L.P. Franca, M. Mallet: A new finite element formulation for computational fluid dynamics. *Comp. Meth. in Appl. Mech. and Eng.* **63** 97-112 (1987).

[119] T.J.R. Hughes, M. Mallet: A new finite element formulation for computational fluid dynamics.*Computer Meth. in Appl. Mech. and Eng.* **54** ,341-355, 1986.

[120] M. Hussaini, T. Zang: Spectral methods in fluid mechanics. *Icase report 86-25* . 1986.

[121] A. Jameson: Transonic flow calculations. In *Numerical methods in fluid mech.* H. Wirz, J. Smolderen eds. p1-87. McGraw-Hill. 1978.

[122] A. Jameson, T.J. Baker: Improvements to the Aircraft Euler Method. *AIAA* 25th aerospace meeting. Reno Jan 1987.

[123] A. Jameson, J. Baker, N. Weatherhill: Calculation of the inviscid transonic flow over a complete aircraft. *AIAA* paper 86-0103, 1986.

[124] C. Johnson: *Numerical solution of PDE by the finite element method* . Cambridge university press, 1987.

[125] C. Johnson: Streamline diffusion methods for problems in fluid mechanics. *Finite elements in fluids* , Vol 6. R. Gallagher ed. Wiley, 1985.

[126] C. Johnson, U. Nävert, J. Pitkäranta: Finite element methods for linear hyperbolic problems. *Computer Meth. In Appl. Mech. Eng* . **45** , 285-312,

1985.

[127] C. Johnson, J. Pitkäranta: An analysis of the discontinuous Galerkin method for a scalar hyperbolic equation. *Math of Comp* . **46** 1-26, 1986.

[128] C. Johnson, A. Szepessy: On the convergence of streamline diffusion finite element methods for hyperbolic conservation laws. in *Numerical methods for compressible flows* , T.E. Tedzuyar ed. AMD-**78** , 1987.

[129] Kato: Non stationary flows of viscous and ideal fluids in R^3 .*J. Func. Anal* . **9** p296-305 ,1962.

[130] M. Kawahara, T. Miwa: Finite element analysis of wave motion. *Int. J. Numer. Methods Eng* . **20** 1193-1210. 1986.

[131] M. Kawahara: On finite element in shallow water long wave flow analysis. in *Finite element in nonlinear mechanics* (J. Oden ed) North-Holland p261-287. 1979.

[132] M. Kawahara: Periodic finite element method of unsteady flow. *Int. J. Meth. in Eng.* **11** 7, p1093-1105. 1977.

[133] N. Kikuchi, T. Torigaki: Adaptive finite element methods in computed aided engineering. *Danmarks Tekniske HØJSKOLE, Matematisk report* 1988-09.

[134] S. Klainerman, A. Majda: Compressible and incompressible fluids. *Comm Pure Appl. Math.* 35, 629-651, 1982.

[135] R. Kohn, C. Morawetz: To appear. See also C. Morawetz: On a weak solution for a transonic flow problem.*Comm. pure and appl. math.* **38** , 797-818,1985.

[136] S.N. Kruzkov: First order quasi-linear equations in several independent variables. *Math USSR Sbornik* , **10** , 217-243, 1970.

[137] Y. A. Kuznetsov: Matrix computational processes in subspaces. in *Computing methods in applied sciences and engineering VI* . R. Glowinski, J.L. Lions ed. North-Holland 1984.

[138] O. Ladyzhenskaya: *The Mathematical Theory of viscous incompressible flows* . Gordon-Breach, 1963.

[139] O.A. Ladyzhenskaya, V.A. Solonnikov, N.N. Ouraltseva: *Equations paraboliques linéaires et quasi-linéaires* . MIR 1967.

[140] H. Lamb: *Hydrodynamics* . Cambridge University Press, 1932.

[141] L. Landau- E. Lifschitz: *Mecanique des Fluides* . MIR Moscou, 1953.

[142] P. Lascaux, R. Theodor: *Analyse numérique matricielle appliquée à l'art de l'ingénieur* . Masson, 1986.

[143] B.E. Launder, D.B. Spalding: *Mathematical models of turbulence* . Academic press, 1972.

[144] P.D. Lax: On the stability of difference approximations to solutions of hyperbolic equations with variable coefficients.*Comm. Pure Appl. Math* . **9** p267, 1961.

[145] P. Lax: *Hyperbolic systems of conservation laws and the mathematical theory of shock waves* , CMBS Regional conference series in applied math. **11** , SIAM, 1972.

[146] J. Leray, C. Schauder: in H. Berestycki, *These d'état* , Université Paris 6, 1982.

[147] A. Leroux: Sur les systèmes hyperboliques. *Thèse d'état.* Paris 1979.

[148] P. Lesaint: Sur la résolution des systèmes hyperboliques du premier ordre par des méthodes d'éléments finis. Thèse de doctorat d'état, Univ. Paris 6 1975.

[149] P. Lesaint, P.A. Raviart: On a finite element method for solving the neutron transport equation. In *Mathematical aspect of finite elements in PDE* . C. de Boor ed. Academic Press, 89-123, 1974.

[150] M. Lesieur: *Turbulence in fluids* . Martinus Nijhoff publishers, 1987.

[151] J. Lighthill: *Waves in fluids* Cambridge University Press,1978.

[152] LINPACK, *User's guide* . J.J. Dongara C.B. Moler, J.R. Bunch, G.W. Steward. SIAM Monograph, 1979.

[153] J.L. Lions: *Quelques Methodes de Resolution des Problèmes aux limites Nonlineaires* . Dunod, 1968.

[154] J.L. Lions. *Controle des systemes gouvernés par des équations aux dérivées partielles* Dunod, 1969.

[155] J.L. Lions, E. Magenes:*Problemes aux limites non homogenes et applications* . Dunod 1968.

[156] J.L. Lions A. Lichnewsky: Super-ordinateurs: évolutions et tendances, *la vie des sciences, comptes rendus de l'académie des sciences,* série générale,1, Octobre 1984.

[157] P.L. Lions: Solution viscosité d'EDP hyperbolique nonlineaire scalaire. Séminaire au Collège de France. Pitman 1989.

[158] R. Lohner: 3D grid generation by the advancing front method. In *Laminar and turbulent flows* . C. Taylor, W.G. Habashi, H. Hafez eds. Pinneridge press, 1987.

[159] R. Lohner, K. Morgan, J. Peraire, O.C. Zienkiewicz: Finite element methods for high speed flows. *AIAA paper* . **85** 1531.

[160] R. Lohner, K. Morgan, J. Peraire, M. Vahdati: Finite element flux-corrected transport (FEM-FCT) for the Euler and Navier-Stokes equations. in *Finite Elements in Fluids* 7 . R. Gallager et al ed. Wiley 1988.

[161] Luo-Hong: Résolution des équations d'Euler en $\Psi - \omega$ en régime transsonique. *Thèse* Université Paris 6. 1988.

[162] D. Luenberger: *Optimization by vector space methods.* Wiley 1969.

[163] R. MacCormack: The influence of CFD on experimental aerospace facilities; a fifteen years projection. *Appendix C* . p79-93. National academic press, Washington D.C. 1983.

[164] Y. Maday, A. Patera: Spectral element methods for the incompressible Navier-Stokes equations. ASME *state of the art survey in comp. mech.* E. Noor ed. 1987.

[165] A. Majda: *Compressible fluid flows and systems of conservation laws* . Applied math sciences series, Springer Vol 53, 1984.

[166] M. Mallet: A finite element method for computational fluid dynamics. Doctoral thesis, University of Stanford, 1985.

[167] M. Mallet, J. Periaux, B. Stoufflet: On fast Euler and Navier-Stokes solvers. Proc. 7th GAMM conf. on *Numerical Methods in Fluid Mechanics.* Louvain. 1987.

[168] A. Matsumura, T. Nishida: The initial value problem for the equations of motion of viscous and heat conductive gases. *J. Math. Kyoto Univ* . **20** , 67-104, 1980.

[169] A. Matsumura, T. Nishida: Initial boundary value problems for the equations of motion in general fluids, in *Computer Methods in applied sciences and Engineering*. R. Glowinski et al. eds; North-Holland, 1982.

[170] F.J. McGrath: Nonstationnary plane flows of viscous and ideal fluids. *Arch.Rat. Mech. Anal* . **27** p328-348, 1968.

[171] D. Mclaughin, G. Papanicolaou, O. Pironneau: Convection of microstructures. *Siam J. Numer. Anal* . **45** ,5, p780-796. 1982.

[172] J.A. Meijerink- H.A. Van der Vorst: An iterative solution method for linear systems of which the coefficient matrix is a symmetric M-matrix. *Math. of Comp* . **31,** 148-162 ,1977.

[173] P. Moin, J. Kim: Large eddy simulation of turbulent channel flow. *J. Fluid. Mech* . **118** , p341, 1982.

[174] K. Morgan, J. Periaux, F. Thomasset: *Analysis of laminar flow over a backward facing step* : Vieweg Vol 9: Notes on numerical fluid mechanics. 1985.

[175] K. Morgan, J. Peraire, R. Lohner: Adaptive finite element flux corrected transport techniques for CFD. in *Finite element, theory and application.* D. Dwoyer, M Hussaini, R. Voigt eds. Springer ICASE/NASA LaRC series. 165-174 1988.

[176] K.W. Morton: FEM for hyperbolic equations. *In Finite Elements in physics:* computer physics reports Vol 6 No 1-6, R. Gruber ed. North Holland. Aout 1987.

[177] K.W. Morton, A. Priestley, E. Suli: Stability of the Lagrange-Galerkin method with non-exact integration. RAIRO-M2AN **22** (4) 1988, p625-654.

[178] F. Murat: l'injection du cone positif de $H^{-1} dans$ $W^{-1,q}$ est compacte pour tout $q < 2.J. Math pures et Appl$. **60** , 309-321, 1981.

[179] J.C. Nedelec: A new family of finite element in R^3. *Numer math* . **50** , 57-82. 1986.

[180] J. Necas: *Ecoulements de fluides; compacité par entropie.* Masson. 1989.

[181] R.A. Nicolaides: Existence uniqueness and approximation for generalized saddle-point problems. *SIAM J. Numer. Anal* . **19** , 2, 349-357 (1981).

[182] J. Nitsche, A. Schatz: On local approximation properties of $L^2 - projection$ on spline-subspaces. *Applicable analysis* , **2** , 161-168 (1972)

[183] J.T. Oden, G. Carey: *Finite elements; mathematical aspects* . Prentice Hall, 1983.

[184] J.T. Oden, T. Strouboulis, Ph. Devloo: Adaptive Finite element methods for high speed compressible flows. in *Finite Elements in Fluids* **7** . R. Gallager et al ed. Wiley 1988.

[185] S. Orszag: Statistical theory of turbulence.(1973). in *Fluid dynamics* . R. Balian-J.L.Peule ed. Gordon-Breach, 234-374, 1977.

[186] B. Palmerio, V. Billet, A. Dervieux, J. Periaux: Self adaptive mesh refinements and finite element methods for solving the Euler equations. *Pro-*

ceeding of the ICFD conf. Reading 1985.

[187] R.L. Panton: *Incompressible flow* . Wiley Interscience,1984.

[188] C. Pares: Un traitement faible par elements finis de la condition de glisse-ment sur une paroi pour les equations de Navier-Stokes. *Note CRAS* . **307** I p101-106. 1988.

[189] R. Peyret, T. Taylor: *Computational methods for fluid flows*. Springer series in computational physics, 1985.

[190] O. Pironneau: On the transport-diffusion algorithm and its applications to the Navier-Stokes equations. *Numer. Math.* **38** , 309-332, 1982.

[191] O. Pironneau: Conditions aux limites sur la pression pour les equations de Navier-Stokes. *Note CRAS* .**303,i** , p403-406. 1986.

[192] O. Pironneau, J. Rappaz: Numerical analysis for compressible isentropic viscous flows. (to appear in IMPACT, 1989)

[193] E. Polak: *Computational methods in optimization.* Academic Press, 1971.

[194] B. Ramaswami, N. Kikuchi: Adaptive finite element method in numerical fluid dynamics. in *Computational methods in flow analysis* . H. Niki, M. Kawahara ed. Okayama University press 52-62. 1988.

[195] G. Raugel: *These* de 3eme cycle, Université de Rennes, 1978.

[196] G. Ritcher: An optimal order error estimate for discontinuous Galerkin methods. Rutgers University, *computer science report* Fev. 1987.

[197] R. Ritchmeyer, K. Morton: *Difference methods for initial value problems.* Wiley 1967.

[198] A. Rizzi, H.Bailey: Split space-marching finite volume method for chemi-cally reacting supersonic flow. AIAA Journal, **14** , 5, 621-628, 1976.

[199] A. Rizzi, H. Viviand eds: *Numerical methods for the computation of in-viscid transonic flows with shock waves* . **3** , Vieweg, 1981.

[200] W. Rodi: Turbulence models and their applications in hydrolics. *Inst. Ass. for Hydrolics.* State of the art paper. Delft. 1980.

[201] G. Rogé: Approximation mixte et acceleration de la convergence pour les équations de Navier-Stokes compressible en elements finis. *Thèse* , Université Paris-6 (1989).

[202] E. Ronquist, A. Patera: Spectral element methods for the unsteady Navier-Stokes equation. *Proc. 7th GAMM conference on Numerical methods in fluid mechanics* . Louvain la neuve, 1987.

[203] L. Rosenhead ed.: *Laminar Boundary layers.* Oxford University Press. 1963.

[204] Ph. Rostand: Kinetic boundary layers, numerical simulations. *Inria report* 728,1987.

[205] Ph. Rostand, B. Stoufflet: TVD schemes to compute compressible viscous flows on unstructure meshes. Proc. 2nd Int Conf. on hyperbolic problems, Aachen (FRG) (To appear in *Vieweg*) (1988).

[206] D. Ruelle: Les attracteurs étranges.*La recherche* , 108, p132 1980.

[207] D. Ruelle: Characteristic exponents for a viscous fluid subjected to time dependant forces.*Com. Math. Phys* . **93** , 285-300, 1984.

[208] I. Ryhming: *Dynamique des fluides* Presses Polytechniques Romandes, 1985.

[209] B. Saramito: Existence de solutions stationnaires pour un problème de fluide compressible. *DPH-PFC* 1224 (1982).

[210] Y. Saad: Krylov subspace methods for solving unsymmetric linear systems. *Math of Comp* . **37** 105-126, 1981.

[211] J.H. Saiac: On the numerical solution of the time-dependant Euler equations for incompressible flow. *Int. J. for Num. Meth. in Fluids* . **5** , 637-656 , 1985.

[212] N. Satofuka: Unconditionally stable explicit method for solving the equations of compressible viscous flows, *Proc. 5th Conf. on numer. meth. in fluid mech* . **7** , Vieweg. 1987.

[213] V. Schumann: subgrid scale model for finite difference simulations of turbulent flows in plane channel and annuli. J. Comp. Physics. **18** 376-404, 1975.

[214] L.R. Scott, M. Vogelius: Norm estimates for a maximal inverse of the divergence operator in spaces of piecewise polynomials. *RAIRO M2AN* . **19** 111-143. 1985.

[215] V.P. Singh: Résolution des équations de Navier-Stokes en éléments finis par des méthodes de décomposition d'opérateurs et des schémas de Runge-Kutta stables. *Thèse de 3ème cycle* , université Pars 6, 1985.

[216] J.S. Smagorinsky: General circulation model of the atmosphere. *Mon. Weather Rev* . **91** 99-164, 1963.

[217] A. Soulaimani, M. Fortin, Y. Ouellet, G. Dhatt, F. Bertrand: Simple continuous pressure element for 2D and 3D incompressible flows. *Comp. Meth. in Appl. Mech. and Eng* . **62** 47-69 (1987).

[218] C. Speziale: On non-linear K-l and $K - \epsilon$ models of turbulence. *J. Fluid Mech.* **178** p459. 1987.

[219] C. Speziale:Turbulence modeling in non-inertial frames of references. ICASE report 88-18.

[220] R. Stenberg: Analysis of mixed finite element methods for the Stokes problem; a unified approach. *Math of Comp* .**42** 9-23. 1984.

[221] B. Stoufflet: Implicite finite element methods for the Euler equations. in *Numerical method for the Euler equations of Fluid dynamics.* F. Angrand ed. SIAM series 1985.

[222] B.Stoufflet, J. Periaux, L. Fezoui, A. Dervieux: Numerical simulations of 3D hypersonic Euler flows around space vehicles using adaptive finite elements *AIAA* paper 8705660, 1987.

[223] G. Strang, G. Fix: *An analysis of the finite element method* . Prentice Hall 1973.

[224] A.H. Stroudl: *Approximate calculation of multiple integrals* . Prentice-Hall 1971.

[225] C. Sulem, P.L Sulem, C. Bardos, U. Frisch: Finite time analycity for the 3D and 3D Kelvin-Helmholtz instability. *Comm. Math. Phys.* , **80** , 485-516, 1981.

[226] E. Suli: Convergence and non-linear stability of the Lagrange-Galerkin method for the Navier-Stokes equations. *Numer. Math.* **53** p459-483, 1988.

[227] T.E. Tezduyar, Y.J. Park, H.A. Deans: Finite element procedures for time dependent convection-diffusion-reaction systems. in *Finite Elements in Fluids* **7** . R. Gallager et al ed. Wiley 1988.

[228] R. Temam:*Theory and Numerical Analysis of the Navier-Stokes eq* . North-Holland 1977.

[229] F. Thomasset: *Implementation of Finite Element Methods for Navier-Stokes eq* . Springer series in Comp. Physics 1981.

[230] N. Ukeguchi, H. Sataka, T. Adachi: On the numerical analysis of compressible flow problems by the modified 'flic' method. Comput. fluids. **8**, 251, 1980.

[231] A. Valli: An existence theorem for compressible viscous flow. *Boll Un. Mat. It. Anal. Funz. Appl* . 18-C, 317-325, 1981.

[232] B. van Leer: Towards the ultimate conservation difference scheme III. *J. Comp. Physics* . **23** , 263-275, 1977.

[233] R. Verfurth: Finite element approximation of incompressible Navier-Stokes equations with slip boundary conditions. *Numer Math* **50** ,697-721,1987.

[234] R. Verfurth: A preconditioned conjugate gradient residual algorithm for the Stokes problem. in R. Braess, W. Hackbush, U. Trottenberg (eds) *Developments in multigrid methods* , Wiley, 1985.

[235] P.L. Viollet: On the modelling of turbulent heat and mass transfers for computations of buoyancy affected flows. *Proc. Int. Conf. Num. Meth. for Laminar and Turbulent Flows* . Venezia, 1981.

[236] A. Wambecq: Rational Runge-Kutta methods for solving systems of ODE, *Computing* , **20** , 333-342, 1978.

[237] L. Wigton, N. Yu, D. Young: GMRES acceleration of computational fluid dynamic codes. *AIAA* paper 85-1494, p67-74. 1985.

[238] P. Woodward, Ph. Colella: The numerical simulation of 2D fluid flows with strong shocks. *J. Comp. Physics* ,**54** , 115-173, 1984.

[239] S.F.Wornom M. M. Hafez: Implicit conservative schemes for the Euler equations. *AIAA Journal* **24** ,2, 215-223, 1986.

[240] N. Yanenko: *The method of fractional steps* . Springer. 1971.

[241] O.C. Zienkiewicz:*The finite element method in engineering science.* McGraw-Hill 1977, Third edition.

[242] W. Zilj: Generalized potential flow theory and direct calculation of velocities applied to the numerical solution of the Navier-Stokes equations. *Int. J. Numer. Meth. in Fluid* **8** , 5, 599-612. 1988.

Appendix

MacFEM:
A finite element software for the Macintosh ^{TM}

1. INTRODUCTION

Numerical analysis is somewhat dry if it is taught without testing the methods. Unfortunately experience shows that a simple finite element solution of a Laplace equation with the P^1 conforming element requires at least 20 hours of programming time; so it is difficult to reach the more interesting applications discussed in this book in the time allotted to a Masters course. On the other hand tests of a ready made subroutine from a library can also be dull and uncreative if the mesh and the data have been prepared before hand.

MacFEM^{TM} is an attempt at a stronger interaction between the program-library and the user; it allows the user to create his data very quickly, thus giving more time to interact with the program. The author's experience is that it is possible to assign projects which involve programing modifications of MacFEM's Fortran routines (Solution of the Euler equation in $\psi - \omega$ is an example).

To reach this aim, we have used fully the graphic interface of modern workstations but since we wanted inexpensive equipment we have used Apple's $Macintosh^{TM}$. The drawback is of course that MacFEM is not portable to other machines; it runs only on any 'Mac' with at least 512K bytes of memory (display of results is in colour on a Mac II).

2. FUNCTIONS

MacFEM is an application program to help the input of data for the finite element methods on triangles in dimension 2. It is menu and mouse driven and allows the user to define graphically the domain Ω and the data of the PDE (right hand side and boundary conditions for example).

- The domain of the PDE Ω is defined by its triangulation which is input entirely with the *mouse* by selecting functions in the *menu* .

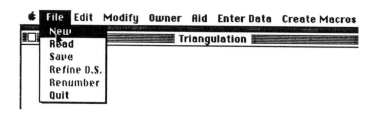

Figure A.1 : *MacFEM 's menu bar with a selection being made (item* **new**) from the *Menu* **File** .

-Boundary conditions are input analytically (ex. $\phi|_\Gamma = x+y-1$). A number i can be attached to any part of the boundary to allow different formulas on different segments. Similarly Ω can be divided into subdomains and a different number j can be attached to each part. The analytical functions that define the right hand side and the boundary condition can also refer to i or j.

This part of MacFEM is written in Pascal and cannot be modified (5000 instructions for the Mac expert because it calls the *Toolbox* .). But the library of programs for the solutions of the PDE by the finite element method is written in FORTRAN and the sources are available to the users (it is necessary to use the Microsoft or Absoft compilers to make modifications to this part).

At the time of writing the following modules are available:

1. Laplace equation with any $\Omega, f, \varphi_\Gamma, g$:

$$-\Delta\varphi = f \text{ in } \Omega, \quad \varphi|_{\Gamma_1} = \varphi_\Gamma, \quad \frac{\partial\varphi}{\partial n}|_{\Gamma_2} = g. \tag{1}$$

Applications :Incompressible inviscid flows (Chapter 2).

Methods : The P^1 finite element method, the linear system is solved by a Cholesky factorization.

2. The transonic equation with any $\Omega, \varphi_\Gamma, g$ in a quasi subsonic regime: $|\nabla\varphi| < 1.2$,

$$\nabla.(1 - \frac{1}{2}|\nabla\varphi|^2)^{2.5}\nabla\varphi = 0 \text{ in } \Omega, \quad \varphi|_{\Gamma_1} = \varphi_\Gamma, \quad \frac{\partial\varphi}{\partial n}|_{\Gamma_2} = g. \tag{2}$$

Applications : compressible potential subsonic and weakly transonic flows (Capture 2)

Method : The P^1 finite element method, the linear system is solved by a Cholesky factorization. Gelder's algorithm with upwinding à la Hughes in the supersonic zones .

3. The convection diffusion equation with any $\Omega, f, \varphi^0, \varphi_\Gamma, g$:

$$\varphi_{,t} + u\nabla\varphi - \nu\Delta\varphi = f \text{ in } \Omega, \ \varphi(x,0) = \varphi^0(x), \quad \varphi|_{\Gamma_1} = \varphi_\Gamma, \quad \frac{\partial\varphi}{\partial n}|_{\Gamma_2} = g. \quad (3)$$

Applications : Convection-Diffusion (chapter 3).
    Methods : The P^1 finite element method and discretisation of the total derivative ($\nu = 0$ is allowed).

4. The Stokes problem in $\psi - \omega$ with any $\Omega, f, \psi_\Gamma, g$:

$$\Delta^2\psi = f \text{ in } \Omega, \psi|_\Gamma = \psi_\Gamma, \quad \frac{\partial\psi}{\partial n} = g \text{ on } \Gamma. \quad (4)$$

Applications : The Stokes problem(chapter 4)
    Methods : The P^1 finite element method on a mixed formulation in ψ, ω; the boundary operator $\omega|_\Gamma \rightarrow \partial\psi/\partial n$ is built and solved by a Choleski factorization (see Glowinski et al.[98] et al. for more details).

5. Solution of the general 2nd order scalar PDE

$$a\varphi - \nabla.(A\nabla\varphi) = f \text{ in } \Omega \quad \varphi|_{\Gamma_1} = \varphi_\Gamma, \quad k\varphi + \frac{\partial\varphi}{\partial n}|_{\Gamma_2} = g, \quad (5)$$

where Ω, $a, A_{11}, A_{12} = A_{21}, A_{22}, f, \varphi_\Gamma, g, k$ are any functions so long as the problem remains strongly elliptic.
    Application : Tool to solve other problems (such as time dependant problems)
    Methods : The P^1 finite element method, the linear system is solved by a Cholesky factorization.

6. The GMRES subroutine, a Stokes equation in P1/P1+bubble and an incompressible Navier-Stokes equations solver are under development.

3. INPUT OF DATA.

3.1 Triangulation
To create a triangulation one can start from pre-triangulated domains and use the following functions
    - Add nodes or triangles
    - Remove nodes, individually or by region
    - Move nodes and glue them to other nodes.
    - Symmetries.

Triangulations are created in two steps; first a sketch is created with few triangles, then it is refined, the scales are defined and the nodes are adjusted to their exact positions as necessary.

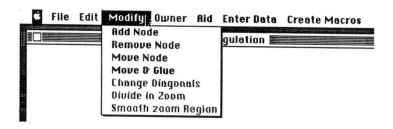

Figure A.2

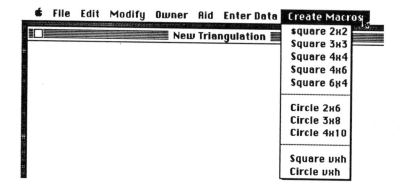

Figure A.3

In the second step one can also:

-refine the mesh by dividing all the triangles of a region into four smaller triangles.

-regularize a region by having each node moved to the center of its neighbours.

-Change diagonals, move, add or remove nodes.

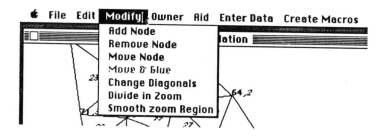

Figure A.4

Triangulation

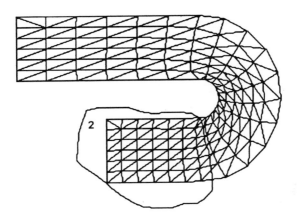

Figure 5 : *Example; This triangulation was obtained by glueing two rectangles and part of a ring and then refining he result; cost of the operations: 2mn. Number 2 is assigned to the region defined by the closed curve*

3.2 Definition of data.

In most practical cases, the data are different on each part of the boundary so it is necessary to number these parts.

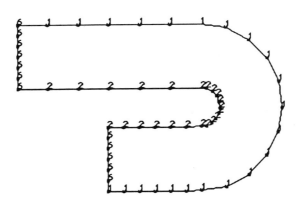

Figure A.6: *To each part of the boundary an integer is assigned by specifying the first and last point of the part of the boundary in the counter clockwise direction.*

Then a compiler of expressions allows the input of each function. Predefined dialogue templates are used for this. The compiler builds arrays of values of the user's functions at the vertices or at the center of gravity of the triangles; these values are stored in a text file for future use.

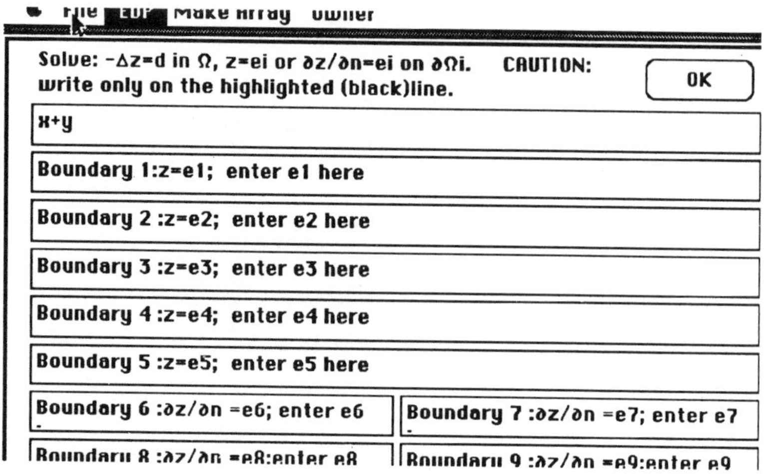

Figure A.7: *The predefined dialogue template to build the two array files f and e made of the values of the right hand side of a Laplace equation (f) and of the values of the boundary conditions on the boundary vertices (e). Here the user has entered f = x + y and the boundary conditions are about to be specified. This takes less than 30 sec.*

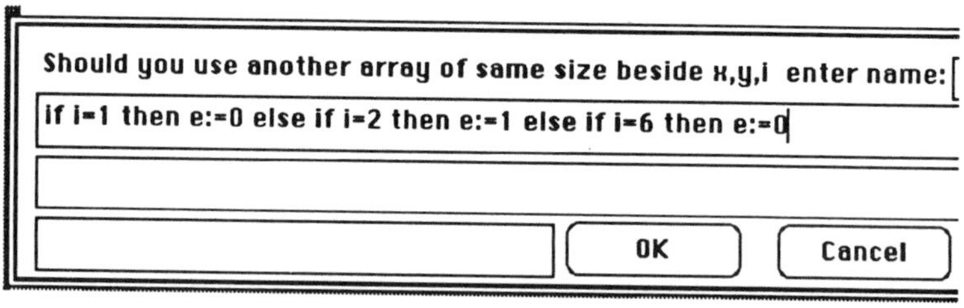

Figure A.8: *Another dialogue template; in this one the user specifies his data by a short Pascal-like instruction.*

3.3 Solutions of the PDEs

The solution part is a separate application (shown below as the first icon) that can be initiated from MacFEM or from the finder. Here, once initiated, PDE-Solver will read the data array e, d prepared by MacFEM and produce an output file array, here *Laplacian* which can be plotted by MacFEM. Note that the solution can be also carried out on another computer also, if the files e and d are transferred; this is easy because they are text files and they can be transferred by any communication program. The file *tube* contains the triangulation data; for Macintosh use it is compacted but it can be stored unpacked if it is to be sent to another computer.

Figure 9: *Various icons of files generated by MacFEM*

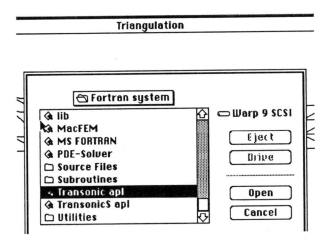

Figure A'.10: *Selection of the resolution module.*

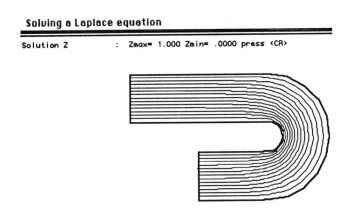

Figure A.11: *Results. In the present case there were 161 nodes and on a MacPlus it took 6 sec.*

3.4 Visualization of results.

The results of the computation are stored in a file; the contours can be plotted (including a zoom on a part of the domain); a plot with colour patches can be made if on a MacII. The results can also be visualized as a curve along a user defined segment.

4. EXAMPLES.

4.1 Compressible flow in a curved pipe

Equation (2)must be solved with $\varphi = 0$ at the inlet, $\rho\partial\varphi/\partial n = 0.4$ on the lateral walls and $\partial\varphi/\partial n = 0.25$ at the outlet (the lower vertical part of the boundary). The flow has a supersonic pocket and the shock can be seen.

RO (Mach 1 is 0.6339:   Zmax= .9680 Zmin= .6687 press <CR>

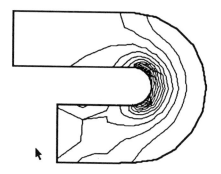

Figure A.12 : *Transonic flow in a curved channel.*

4.2 Transonic flow with shocks.

Again (2) must be solved with the same boundary conditions except at the outlet where $g = 0.25$ and on the following geometry (nozzle):

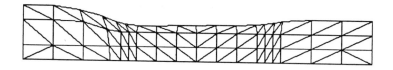

Figure A.13 : *Triangulation; symmetry is used to reduce time.*

RO (Mach 1 is 0.6339:   Zmax= .8709 Zmin= .4599 press <CR>

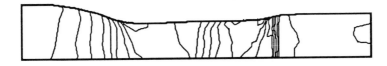

Figure A.14 :*Results; the supersonic zone and the shock can be seen.*

Upwinding is cames out with values on the upstream triangles. However Gelder's algorithm is not very stable in this case and there seem to be two subsequences each converging to a different limit. We plan to improve this computatio